D1498264

THE CREATION OF
QUANTUM CHROMODYNAMICS
AND THE EFFECTIVE ENERGY

World Scientific Series in 20th Century Physics

World Scientific Series in 20th Century Physics – Vol. 25

THE CREATION OF QUANTUM CHROMODYNAMICS AND THE EFFECTIVE ENERGY

IN HONOUR OF A. ZICHICHI ON THE OCCASION OF THE GALVANI BICENTENARY CELEBRATIONS

V.N. Gribov

G. 't Hooft

G. Veneziano

V.F. Weisskopf

Edited by

L.N. Lipatov

 World Scientific
Singapore • New Jersey • London • Hong Kong

Published by

World Scientific Publishing Co. Pte. Ltd.

P O Box 128, Farrer Road, Singapore 912805

USA office: Suite 1B, 1060 Main Street, River Edge, NJ 07661

UK office: 57 Shelton Street, Covent Garden, London WC2H 9HE

Library of Congress Cataloging-in-Publication Data
The creation of quantum chromodynamics and the effective energy : in honour of A. Zichichi on the occasion of the Galvani bicentenary celebrations / V.N. Gribov ... [et al.] ; edited by L.N. Lipatov.
    p. cm. -- (World Scientific series in 20th century physics ; vol. 25)
  ISBN 9810241410
    1. Quantum chromodynamics. 2. Effective interactions (Nuclear physics) I. Zichichi, Antonino. II. Gribov, V.N. (Vladimir N.) III. Lipatov, L.N. (Lev Nikolaevich), 1943-
IV. Series.

QC793.3.Q35.C74 2000
539.7'548--dc21   00-040870

British Library of Cataloging-in-Publication Data
A catalogue record for this book is available from the British Library.

The editor and publisher would like to thank the following publishers for their assistance and their permission to reproduce the articles found in this volume:

Elsevier Science Publishers B. V. (*Physics Letters*), Plenum Publishing Corporation, Società Italiana di Fisica (*Il Nuovo Cimento, Lettere al Nuovo Cimento*)

First edition jointly published in 1997 by Accademia Delle Scienze Bologna, Alma Mater Studiorum, Istituto Nazionale di Fisica Nucleare and Società Italiana di Fisica.

Printed in Singapore by FuIsland Offset Printing

This volume is dedicated to the experimental contributions in building QCD performed at CERN by A. Zichichi and his group during two decades from the middle 60s to the middle 80s, and especially to the Effective Energy and the Universality Features discovered in QCD processes.

CONTENTS

PREFACE

The University of Bologna and its Academy of Sciences, in collaboration with the Italian National Institute for Nuclear Physics (INFN) and the Italian Physical Society (SIF), celebrate this year the bicentenary of a great pioneer in the field of electric phenomena, Luigi Galvani, the father of macroelectricity.

During these two centuries, the physics of electric phenomena has given rise first to the Maxwell equations, then to Quantum Electrodynamics, and finally to the synthesis of all reproducible phenomena, the "Standard Model".

A cornerstone of the Standard Model is Quantum ChromoDynamics (QCD) which describes the interaction between quarks and gluons in the innermost part of the structure of matter.

The discovery of QCD will be recalled in the future as one of the greatest achievements of mankind. Many physicists — the world over — have contributed to its creation on both the experimental and the theoretical front.

We are proud that one of our most distinguished colleagues, Professor Antonino Zichichi, has played an important role in this scientific venture, as documented by his works which are reproduced in the present volume.

One of the founders of European Physics, Professor Victor F. Weisskopf, contributes with his memories of the time when QCD had many problems, especially three to be confronted with. This volume owes its existence to a founding father of QCD, Professor Vladimir N. Gribov, whose sudden loss prevented him from directly contributing to its final edition. He wanted to emphasise the value of experimental discoveries in the field of Subnuclear Physics, an experimental science which constantly needs to be confronted with original theoretical ideas. Two world leaders in Subnuclear Theoretical Physics, Professors Gerardus 't Hooft and Gabriele Veneziano, illustrate the significance and the value of the contributions of Antonino Zichichi in QCD.

This volume is a joint publication of the University of Bologna with its Academy of Sciences, INFN and SIF for the Galvani Bicentenary Celebrations. The volume is edited by one of the closest collaborators of Professor Vladimir N. Gribov: Professor Lev N. Lipatov. For his most appreciated collaboration we would like to thank the editor, who is one of the most original contributors to the field of QCD, in both its perturbative and nonperturbative aspects.

Ottavio Barnabei
President of the Bologna Academy of Sciences

Enzo Iarocci
President of the Italian Institute for Nuclear Physics

Renato Angelo Ricci
President of the Italian Physical Society

Fabio Roversi Monaco
Rector of the University of Bologna

Bologna 1998 Galvani Bicentenary Celebrations

«The agreement between the momentum distributions obtained in e^+e^- annihilation and in pp collisions suggests that the mechanism for transforming energy into particles in these two processes, so far considered very different, must be the same.» From: **Evidence of the same multiparticle production mechanism in p-p collisions as in e^+e^- annihilation**, M. Basile, G. Cara Romeo, L. Cifarelli, A. Contin, G. D'Alì, P. Di Cesare, B. Esposito, P. Giusti, T. Massam, F. Palmonari, G. Sartorelli, G. Valenti and A. Zichichi, *Physics Letters* <u>92B</u>, 367 (1980).

This paper, the first of the series, is the result of a detailed study of many thousands of pp collisions, each one measured in detail. The nominal energy $(\sqrt{s})_{pp}$ was 62 GeV, but the "effective" energies covered a continuous spectrum. The understanding of the continuous spectrum was the key to the experimental discovery. In fact (pp) interactions do produce a continuous set of multihadronic final states identical to those produced in (e^+e^-) annihilation at the same "effective" energies.

INTRODUCTION

Lev N. Lipatov

St. Petersburg Nuclear Physics Institute (PNPI), Gatchina, St. Petersburg, Russia

INTRODUCTION

Lev N. Lipatov

St. Petersburg Nuclear Physics Institute (PNPI), Gatchina, St. Petersburg, Russia

Quantum-Chromo-Dynamics (QCD) is an essential part of the Standard Model of elementary particle interactions. It describes the strong interaction of hadrons constructed from quarks and gluons. The gluons are quanta of the self-interacting Yang-Mills field responsible for colour forces between quarks. This remarkable theory was invented more than 25 years ago as a result of many experimental and theoretical discoveries including the explanation of the scaling behaviour of structure functions for the deep-inelastic *ep* scattering. Some of the ingredients of QCD are clarified in this volume.

In the paper of Victor F. Weisskopf three physical problems that were a challenge for QCD are discussed. They were investigated experimentally by Antonino Zichichi and his group during two decades of activity at CERN.

In the report by Gerardus 't Hooft a detailed experimental and theoretical history of QCD is given. The three problems considered by V.F. Weisskopf (U(1)-anomaly, quark confinement and effective energy) are placed by G. 't Hooft in the general context of QCD and of its development.

The contribution of Gabriele Veneziano is entirely focused on the problem of the Effective Energy which is related to the discovery at CERN in

1980, that in pp interactions at ISR the multi-hadronic final states have properties similar to those produced in e^+e^- annihilation, provided that the Effective Energy, introduced by A. Zichichi, is the same.

Before this discovery Vladimir N. Gribov had stressed that the high energy behaviour of cross-sections for e^+e^- annihilation and pp scattering should be different because in the second case, contrary to the first one, the number of the s-channel partial waves grows with energy. From this point of view the similarity of the hadron distribution in the final states of these reactions looks surprising.

In the framework of the QCD parton model, such similarity could be related with the universality of the fragmentation functions describing the transition of quarks into hadrons. In the leading logarithmic approximation (LLA), the fragmentation functions and the parton inclusive probabilities coincide according to the Gribov-Lipatov reciprocal relation. In the next-to-leading approximation for the splitting kernels of the Dokshitzer-Gribov-Lipatov-Altarelli-Parisi (DGLAP) equation, this relation is violated due to the Sudakov effect of the soft gluon emission but, according to the Fadin-Ermolaev-Mueller coherence phenomenon, the fragmentation functions in the small-x region satisfy a simple evolution equation corresponding to the angular ordering of the emitted gluons.

In high energy hadron scattering processes there is, apart from the mechanism of the jet hadronization, also a multi-peripheral contribution to the multi-particle production. By means of the optical theorem, this contribution is related to the Pomeron exchange in the elastic scattering amplitude. In QCD the Pomeron in LLA is a composite state of two dressed gluons. Its wave function satisfies the Balitsky-Fadin-Kuraev-Lipatov (BFKL) equation. Recently the next-to-leading corrections to this equation were calculated, which give a possibility to fix the region of applicability of the BFKL dynamics. These corrections determine the relative rapidity interval between two produced jets below which their correlation becomes essential. S-matrix unitarization effects are related with the multi-Pomeron exchanges. In LLA of the multi-colour QCD, the equations for the composite states of several gluons have remarkable mathematical properties leading to their integrability. The distribution of the produced jets in the deep-inelastic *ep* scattering can discriminate DGLAP and BFKL dynamics.

The value of introducing the "Effective Energy" is illustrated well by Gribov's enthusiastic comments made after publication of the paper of Zichichi and collaborators in 1980. This important paper gave rise to the "Universality Features" discovered later in all QCD induced processes.

Of the three difficulties described by Weisskopf, the U(1) anomaly and confinement problems have now a satisfactory theoretical solution, whereas the Effective Energy and the Universality Features have not.

In fact, as emphasised by V. Gribov, in order to be able to "predict" Universality Features, QCD needs to be formulated in such a way that both perturbative and non-perturbative effects are included simultaneously. This is the case where the experimental results precede theory as explained by G. 't Hooft and G. Veneziano in their reports.

For this reason, a selected sample of papers by A. Zichichi and collaborators, dealing with the introduction of the "Effective Energy" and with the subsequent discovery of the "Universality Features" in QCD induced processes, is reproduced in this book.

The work of Zichichi in QCD started with his experimental contributions to: i) the confinement property of QCD via his ISR (*pp*) experiment to detect free quarks; ii) the ($\eta\eta'$) problem, i.e. the U(1) anomaly via his detection of the $X^0 \to \gamma\gamma$ decay mode. It was culminated with iii) his 1980 ISR discovery paper of fundamental importance for the validity of QCD. Not because QCD is able to "predict" the Universality Features as emphasised by Gerardus 't Hooft and Gabriele Veneziano in their reports, but because the Universality Features must exist if quarks and gluons are the basic ingredients of all hadronization processes.

All in all, these remarkable papers are of great interest because, as emphasized by Gabriele Veneziano in his report: «*they represent the "hidden" side of QCD physics, which no-one knew (and even now knows) how to reduce to a common basis. The concept of "Effective Energy" was needed, but this quantity was not simple to disentangle in the analysis of multihadronic final states. The first paper of this series [1] was in fact the result of a detailed study of many thousands of pp collisions, each one measured in detail. The total energy in the pp collisions was 62 GeV, but the effective energy covered a continuous spectrum*». The understanding of this continuous spectrum was the key of the experimental discovery.

The content of this volume should be interesting for all physicists working in the field of elementary particles who want to know the history of QCD and the status of the difficult problems of hadrons interactions.

THREE PROBLEMS FACING QCD

Victor F. Weisskopf

MIT, Cambridge, USA

THREE PROBLEMS FACING QCD

Victor F. Weisskopf
MIT, Cambridge, USA

One of my contributions to CERN was the decision to construct the first proton-proton collider of the world, the Intersecting Storage Ring (ISR). The ISR allowed to reach the highest collision energy in the interaction between protons. I am therefore very happy that it is thanks to a series of experiments performed by A. Zichichi and collaborators with the ISR that the multitude of final states produced when particles interact strongly electromagnetically and weakly can be put on the same basis.

Let me try to report my personal recollection of how the "Effective Energy" entered into the physics of strong interactions, which started in 1947 with the discovery of the π–meson and soon became the physics of very many strongly interacting particles, baryons and mesons, whose existence had no theoretical prediction whatsoever.

The physics of strong interactions whose "elementary particles" were of baryonic and mesonic nature was characterised by two classes of phenomena, both showing a very large number of different varieties. One was of static nature: i.e. the enormous number of mesons and baryons. The other was of dynamic nature: the enormous number of different final states produced by different pairs of interacting particles, such as (πp), (pp), (p̄p), (Kp) in addition to (e⁺e⁻), (νp), (μp), (ep) etc.. These multihadronic final states were all measured to be different.

What is now called Quantum Chromodynamics (QCD), the non-Abelian force acting between quarks and gluons, is the result of an impressive amount of work by experimentalists and theoreticians, all contributing to disentangle the various obstacles.

Three of these were: i) the apparent contradiction between scaling in Deep Inelastic Scattering (DIS) discovered at SLAC, and the impossibility for the protons to break up into their constituents i.e. the confinement problem; is confinement understood as a rigorous consequence of QCD? ii) the heaviness of the pseudoscalar mesons, better known as the U(1) anomaly; in other words, do the η and η' behave as they should in QCD? iii) the large variety of multihadronic final states characterizing each pair of interacting particles, i.e. the lack of Universality Features qualitatively expected by QCD but never observed in QCD processes. Why do the multihadronic final states depend on the nature of the colliding pair?

Nino contributed to all of them with his skill and ingenuity. Especially important are his contributions to solve the third problem. This is the reason why this volume is also dedicated to the reproduction of the most important papers, including the detailed references to the other works in the same field, dealing with the "Effective Energy" and the Universalities Features discovered at the ISR studying (pp) interactions.

Let me go back with my recollections related to the first QCD obstacle.

When the quark proposal by Gell-Mann and Zweig came in 1963, the theoretical front was dominated by S-Matrix Theory, which meant the abandon of the basic principles of Relativistic Quantum Field Theory (RQFT). The idea that quarks could be the constituents of baryons and mesons allowed the "eightfold way" to have its fundamental representation: the quark triplet (u, d, s). It took nevertheless ten years before Fritzsch, Gell-Mann and Leutwyler could imagine the existence of a new degree of freedom, colour, for quarks and gluons.

During these ten years, other experimental results obtained in strong interactions gave rise to a strong support for the S-Matrix Theory and the discovery of scaling at SLAC appeared to confirm the view that no RQFT could account for all these findings.

In fact scaling suggested that the "partons" inside a proton behave as "free" non-interacting constituents. And this was believed not to be expected by any RQFT. Morever free quarks had never been observed. Nino and his group had been searching for free quarks in the cosmic radiation. But the SLAC discovery was a very high energy process. Thus, in the same year when scaling was discovered in DIS at SLAC, Nino started at CERN a project which allowed in the early 70s to implement an experiment with the highest energy (pp) collider of the world, the ISR. If scaling were true, two colliding protons should break into their constituents. If among these constituents there were quarks, the ISR should have produced fractionally charged particles. These were not observed despite the very high energy of the (pp) collisions. Thus "confinement" was established on firm grounds. And "confinement" became another serious problem for QCD, in addition to the well known one concerning the heaviness of the pseudoscalar mesons.

The experimental contributions of A. Zichichi and collaborators to this second QCD obstacle, also known as the $(\eta-\eta')$ problem or the U(1) anomaly, are the discovery of the decay into 2γ of the X^0-meson (now called η') and the studies of the radiative decays of the heavy pseudoscalar mesons. This series of experiments made use of what Nino called an "NBC" (non Bubble-Chamber) set up. These were in fact years dominated by the Bubble-Chamber technology but Nino had decided that the interesting issues were elsewhere and new non Bubble-Chamber devices were needed.

The ISR (pp) experiment producing no quarks and the scaling effect discovered at SLAC are no longer in contradiction, but now explained by QCD in terms of confinement and of asymptotic freedom. And confinement is finally in a satisfactory theoretical shape.

The very heavy mass of the X^0 (now called η') is understood, together with the mass of the η and of the π-meson, in terms of Instantons. A great theoretical step.

Let me turn to the third obstacle: why do the multihadronic final states appear to be all different?

The "static" multitude of hadronic states could be reduced by an order of magnitude, in terms of octets and decuplets, thanks to the "Eightfold way" of Gell-Mann and Ne'eman. But the "dynamical" multitude of the different final states was far from showing any such hint.

In particular, the fact that each pair of hadronic particles (πp), (pp), $(\bar{p}p)$, (Kp) etc., was producing its own final state, with different properties, such as the average charged multiplicity, the fractional momentum distribution of the secondary particles, and so on, was very disturbing. In fact, the properties of these final states appeared to depend on the nature of the pair of interacting "elementary" particles. This was called by Gribov the "hidden" side of QCD.

Here QCD had no quantitative predictions but qualitative ones. If all baryons and mesons are made of the same constituents, quarks and gluons, then the multihadronic final states should show Universality Features. It is exactly here that Nino succeeded to solve the problem with his new quantity entering in all processes: the "Effective Energy". He had the right idea at the right time. In fact, the Universality Features are experimentally hidden insofar as the "Effective Energy" is not introduced in the analysis.

It is for me very gratifying that the discovery of the "Effective Energy" was implemented at the CERN-ISR. Moreover, I like it because it still has to be quantitatively explained by QCD.

THE CREATION OF
QUANTUM CHROMODYNAMICS

Gerardus 't Hooft

Institute for Theoretical Physics, Utrecht University, The Netherlands

THE CREATION OF QUANTUM CHROMODYNAMICS

Gerardus 't Hooft

Institute for Theoretical Physics, Utrecht University
Princetonplein 5, 3584 CC Utrecht, The Netherlands
email: g.thooft@fys.ruu.nl

April 3, 1998

ABSTRACT

This is a personal account of those events which, in the eyes of the author, were the most crucial ones in the development of our insights in the nature of the strong interactions.

1   PREHISTORY

In the early '60s, most physicists did not believe that strong interactions were described by a conventional quantum field theory. Much effort went into the attempts to axiomise the doctrine of relativistic quantum particles strongly interacting among themselves. All we had, for sure, was that there should exist an S-matrix, obeying all sorts of requirements: unitarity, causality, crossing symmetry [1] [2].

One then turned to the experimental evidence that strong interactions have a well-documented symmetry structure. There is a practically exact $U(2)$

symmetry (isospin and baryon number conservation), a softly broken $SU(3)$ (M. Gell-Mann's "Eightfold way" [3]), and a more delicate chiral $SU(2) \times SU(2)$ that is both spontaneously broken (so that the pions became massless Goldstone bosons) and explicitly broken (the pions still have a non-vanishing mass). Although the possibility of a substructure for the strongly interacting particles — the quark theory [4] — was soon realised, it was nevertheless suspected that not the quarks but the Noether currents associated with the symmetries were playing an essential role in the dynamics. It was attempted to write down rigorous algebras for these currents, such that their representations would generate the sought-for S-matrix [5]. They were combined with unitarity, causality and crossing symmetry as these were suggested by quantum field theory, but one did not want to use quantum field theory itself because of the infinities generated by the inevitable renormalization procedure [6].

A promising new idea was "duality" (in terms of the Mandelstam variables s, t and u for elastic scattering events) [7]. A great discovery by G. Veneziano [8] was that one could write down simple mathematical expressions for amplitudes that obeyed this duality requirement. Since this duality is somewhat different from what one has in standard quantum field theory, it was clear why there was much enthusiasm for it. After Z. Koba and H.-B. Nielsen found how to generalise Veneziano's expressions to n-particle amplitudes [9], the physical interpretation was also discovered, by Nielsen [10], T. Goto, L. Susskind and Y. Nambu [11]: the amplitudes are the scattering amplitudes for strings colliding, merging and splitting one against the other.

However, these string theories were confronted with difficulties concerning unitarity, which appeared to require the presence of tachyons. The latter are physically not observed, and furthermore would lead to technical inconsistencies in the theory. Later, it was found how to safeguard string theories against this disease, but then they would produce a gravitational force, and require supersymmetry. This made these theories more suitable for Planck length physics, but it disqualified them for the "ordinary" strong interactions. We now know that the "string theory for the strong interactions" is only approximately true; in QCD, color fields tend to form vortices that behave similarly to strings, and this explains the early qualitative successes of the dual models.

What were the alternatives? In 1960, Gell-Mann and M. Lévy [12] had constructed a simple renormalizable model for the strong interactions, the so-called 'sigma-model'[1], in which the nucleons, p and n, and the pions, $\pi^\pm$ and π^0 played a dominant role, together with a new hypothetical particle, the σ. In this theory, chiral $SU(2) \times SU(2)$ was spontaneously and explicitly broken much in the way actually observed, and some of the features that had been derived from the algebra of the chiral currents could be explicitly demonstrated in this model, which made it very attractive. This produced, in 1969, a short revival of quantum field theory. It was extensively studied by B.W. Lee [13], J.-L. Gervais [14] and K. Symanzik [15][2]. Yet, the model contained one very large coupling constant, the pion-nucleon coupling constant $g \approx 13$, which made it clear that ordinary perturbation expansions make no sense.

There were hopes that perturbation expansion could be improved. The

---

1   In mathematical physics, sigma models are popular subjects of investigation; the origin of the name is usually forgotten.

2   These authors were certainly aware of the fact that the explicit chiral symmetry breaking could be attributed to the mass terms of the u- and the d- quark, even if the details of the quark dynamics were not understood.

magic word[3] was 'Padé approximation', a rearrangement of the perturbative corrections in terms of rational functions of the perturbation parameter g. Successes were reported in obtaining phenomenologically reasonable expressions, but it became clear that this procedure could not serve as a theoretically sound framework.

Meanwhile, experiments were moving towards the higher energy domain, and here a peculiar scaling phenomenon was observed, and the general nature of this behaviour was investigated by J.D. Bjorken [17]. This experimental discovery became to be known as "Bjorken scaling". It was subsequently interpreted by Feynman [18] as if, at short distances, constituent particles inside hadrons move comparatively freely. He called these constituents 'partons'. It would later turn out to be a good insight *not* to call them 'quarks', because now we know that partons can be both quarks and gluons.

The study of scaling behaviour of renormalizable theories became a central point around 1970. Independently, in the late '60s, C.G. Callan [19] and K. Symanzik [20] found equations obeyed by the renormalized Green functions that follow from their scaling behaviour. By inspecting the Lagrangian of a field theory, and the ensuing classical field equations, one might have expected that the behaviour of such theories under scale transformations is quite simple, especially in cases where one can ignore the effects from the particle mass terms. But these authors found that the quantum corrections to scaling behaviour are actually quite subtle: the naive scaling equations are violated by the radiative corrections. It should be stressed that these radiative corrections depend on many details of the model under consideration, and Callan and

---

3   The need for "magic" is a deplorable weakness in any theory [16].

Symanzik only studied Quantum Electrodynamics (an Abelian gauge theory) and theories involving scalar particles, not the non-Abelian gauge theories. Functions in their equations they called β-functions, describe how the effective coupling constants actually depend on the scale. This scale dependence had been found earlier by Gell-Mann and F. Low [21]. The fact that the redefinitions of the renormalized coupling constants form a group had been emphasized earlier by Peterman and Stueckelberg [22], hence the name 'renormalization group', but only the subgroup that corresponds to scale transformations has physically significant consequences.

Callan and Symanzik explicitly constructed their equations only for the theories mentioned earlier, but they expected them to have universal validity. In particular, the β-functions were found to be positively valued, and this too was thought to be a universal feature, valid for *all* quantum field theories, Abelian as well as non-Abelian ones. There were mathematical arguments supporting this view, arguments which were only later seen to be flawed.

In the late '60s, there were various developments in the domain of Yang-Mills theories. Models for the electroweak interactions had been proposed by Glashow [23], Salam [24], and Weinberg [25], but they did not know how these models should be renormalized while avoiding the complications due to ghost particles with 'negative metric', which means that these violate unitarity. Consequently, these investigators turned their attention elsewhere.

After some pioneering work by R.P. Feynman [26] and B.S. DeWitt [27], the investigation of pure Yang-Mills theories was pursued by a group in Leningrad (now St. Petersburg), L. Faddeev and V. Popov [28], and in Moscow, E.S. Fradkin and I.V. Tyutin [29], although these authors were primarily interested in the closely related, but much more difficult problem of quantizing gravity. Yang-Mills gauge theory with a mass term explicitly added (without Higgs mechanism) had been introduced by Feynman [26] and Glashow [23], and this avenue was pursued by Veltman [30].

In their studies of the Gell-Mann-Lévy σ-model, it was observed, by B.W. Lee and Symanzik, that renormalization and symmetry breakdown went well together. As a student at that time, I was interested in applying this observation to what is now called the Higgs mechanism. The Higgs mechanism differs in several important respects from spontaneous breakdown of a global symmetry. Firstly, the Higgs effect alters in an essential way the particle spectrum, as it removes scalar particle modes while adding the physical transverse helicity states of the gauge vector particles. Secondly, it adds various kinds of "ghost particles" to the Feynman rules of the system. Thirdly, the implications of local gauge invariance were ill understood, and the exact prescriptions for higher loop diagrams were unknown. Veltman had pointed out to me that scalar particles can only exacerbate the unitarity problems of massive vector particles, if you try to use them to remove the $k_\mu k_\nu$ terms in the vector propagator. Yet this was when you used derivative couplings, instead of applying gauge invariance. My suspicion was that nevertheless the Higgs mechanism should be essential for renormalization, and it appeared to allow for a treatment quite similar to that of the σ-model within perturbation theory. Discussions with Lee and Symanzik were not helpful at that time, as they both confessed not to know much about Yang-Mills theories. My advisor Veltman did not want to hear about fields with vacuum expectation values.

Realizing that the world was not yet ready to accept the renormalizable Higgs theories, I returned to the case of pure, massless Yang-Mills fields. At that time, renormalization was generally thought to be a trick that happened to work only in very special cases such as Quantum Electrodynamics. In 1970, the details of the renormalization procedure for pure Non-Abelian gauge theories were worked out [31]. In this paper, a new set of extended Ward identities was derived, and their essential significance for the renormalizability was proved[4].

Generalization towards gauge theories with Higgs mechanism was now straightforward. A key to the understanding of the renormalizability of theories with massive gauge bosons was the realization that mass terms should *never* be inserted by hand, but exclusively evoked by a Higgs mechanism. I now could convince Veltman of this fact, by providing the exact Feynman rules and the Ward identities needed for renormalization [34]. These renormalization procedures were immediately extended to all Abelian and non-Abelian gauge theories, with or without Higgs mechanism [34][5]. Numerous investigations were carried out to further streamline the renormalization procedures, among which the introduction of dimensional regularization and renormalization [35] and BRST quantization [33].

Only after the difficulties of the massless and massive Yang-Mills theories were cleared, did theoreticians have the tools in their hands to construct theories such as Quantum Chromodynamics.

2   AWAKENING

In the '60s, various attempts were made to construct Yang-Mills theories for the strong interactions. Models were invented in which the vector mesons

---

4   These identities were later to be called Slavnov-Taylor identities [32], because these authors extended the identities to off-shell amplitudes. The only reason why the off-shell expressions were not written down earlier was that they create new infinities, requiring the renormalization of wave functions and composite operators, while these were not directly needed for the renormalization of the Yang-Mills fields (for the massless as well as the Higgs case). The identities were later understood to be a consequence of an induced anticommutative symmetry by Becchi, Rouet and Stora, and by Tyutin [33].

5   Provided that some chiral anomalies are made to cancel out.

were treated as Yang-Mills bosons, with a mass term added to the Yang-Mills action. This was not a bad idea, because it explained a phenomenon called vector meson dominance, an experimentally observed feature of the electromagnetic form factors of the nucleon [36]. Of course, at that time, it was not known how to renormalize these models. We now know that they are not renormalizable[6].

What was really needed, however, was a model that described quarks. Ever since Gell-Mann had launched the 'Eightfold Way', the idea that hadrons are made up of quark-like constituents [4] was kept alive. The big question was how these quarks are bound together. Gell-Mann knew about Yang-Mills theory. H. Fritzsch, who had just escaped from the DDR, joined him in his research; he too had studied Yang-Mills theory. They emphasized that, if you attach a color degree of freedom to quarks, you remove an apparent contradiction arising from the Fermi-Dirac statistics of quarks on the one hand, and the observed $SU(3)$ symmetric representation spectrum of baryons on the other [37]. All that was needed now was a confining two-body or three-body force that selects out color singlet states for the hadrons, rendering all color-non-singlets unphysical. Exactly why they are unphysical (do such states carry infinite energy, or is it enough to say that they are not gauge-invariant?) was unclear [38].

The notion of color as an internal symmetry for quarks had been introduced by O.W. Greenberg [39], and it was used by M.Y. Han and Nambu [40] in their model for the strong interactions. They avoided the quark confinement problem by invoking a mixing mechanism for color and

---

6   They could be cured, by invoking Higgs mechanisms, but these effective strong
     interaction theories suffer from other severe defects as well.

electromagnetism in such a way that quarks receive integral charges. This idea is not as far off from the truth as one might think. If, in modern QCD, scalar fields would exist in the fundamental representation of $SU(3)$, regardless whether they possess a vacuum expectation value or not[7], one would be able to treat these as Higgs fields, to represent the situation in the way Han and Nambu did [41].

Gell-Mann, Fritzsch and H. Leutwyler proposed an unbroken $SU(3)$ Yang-Mills theory for quarks [38]. It was not clear to them whether or to what extent one had to modify the Yang-Mills potential in order to achieve quark confinement and Bjorken scaling. They thought of non-local terms in the action, but they also suggested that complicated dynamical mechanisms could exist that take care of the problem.

In spite of their beauty, these ideas were taken with skepticism. The explanation of quark confinement was not considered to be adequate. This skepticism encouraged a new series of experiments at CERN, where the highest energy pp collider, ISR, allowed to check if quarks remained confined at these energies [42]. It seemed to be a miracle that quarks can move freely at high energies and yet they stick together inseparably inside the hadrons. To many physicists, Bjorken scaling appeared to be a signal that no quantum field theory would be fit to account for the observed strong interactions [43]. Why then did nobody compute the β-function for Yang-Mills theories? I *had* computed this function, and knew about its sign. In 1971, as soon as we knew how to renormalize the theory, I started a preliminary calculation, showing the negative sign. Not realizing that this was something totally novel, I could not imagine

7   This would not be a gauge-invariant statement anyway.

that this was all that was needed to resolve the outstanding problems of strong interaction theory. I actually thought that this scaling behaviour was also known, but not considered relevant for partons. Since I could not understand their problems, I turned away from the strong interactions that had to be infinitely complicated.

By 1972, when Veltman and I had worked out many examples of gauge models, we knew how to compute the renormalization counter terms, and I had found a way to relate these to the equations governing the running coupling constants. Veltman was not interested in these findings. According to him, they were only relevant for Euclidean space, which has nothing to do with experimental observations, all done in Minkowski space. On several occasions I mentioned to him the possibility of a pure gauge theory for the strong interactions, but his mind was made up. As long as they produced no answer to the quark confinement problem, such ideas are useless, and no referee would accept a publication suggesting such an idea.

In June 1972, a small meeting was organised in Marseille by C.P. Korthals Altes. Symanzik was there. Upon arrival, right at the airport, Symanzik and I began to discuss our work. Symanzik explained that he worked on $\lambda\phi^4$ theory with negative λ, because of its interesting scaling properties [44]. He had been attracted by the idea that theories with a negative β-function would reproduce parton-like behaviour, but, the only models with such properties that he could think of were theories with a negative squared coupling constant, $g^2 < 0$ or $\lambda < 0$.

Naturally, I asked him how he would solve the problem that the energy appears to be unbounded from below. He explained that perhaps this problem could be cured by non-perturbative effects, which at that time were very badly understood. But I informed him of my finding that the coefficient determining the running of the coupling strength, that he called $\beta\,(g^2)$, for gauge theories is negative. Symanzik was surprised and skeptical. "If this is true, it will be very important, and you should publish this result quickly, and if you won't, somebody else will," he said.

In the meeting, Symanzik ended his talk on negative $\lambda\phi^4$ theories by the remark that further study on the β-functions for non-Abelian gauge theories were called for. When it was time for questions I came forward to write down on the blackboard the expression I had derived:

$$\frac{\mu dg^2}{d\mu} = \frac{g^4}{8\pi^2}\left(\frac{-11}{3}C_1 + \frac{2}{3}C_2 + \frac{1}{6}C_3\right),$$

for a theory in which the gauge group has a Casimir operator C_1 and where fermions have a Casimir operator C_2. Scalars contribute with the Casimir operator C_3. In the $SU(2)$ case this would mean that with less than 11 fermions in the elementary representation (and no scalars) the β-function would be negative, and for $SU(3)$ the critical number would be 33/2.

Why did I not follow Symanzik's sensible advice to publish this result? Weinberg had written a long paper in which he reported about the complicated infrared structure of gauge theories, which made me believe he knew everything about what was to be called 'infrared slavery', a direct consequence of 'asymptotic freedom'. Surely, I thought, he understands the scaling behaviour of gauge theories. David Gross had explained to me that Bjorken scaling cannot be accommodated for in any field theory, Abelian or not. Veltman was telling me that scaling all components of the momentum vector would lead you far away from the mass shell, whereas *all* experiments were looking at on-mass-shell hadronic final states, so the deep-Euclidean region was unphysical and uninteresting[8]. As for my views about the small-distance structure of quantum field theories, I felt like crying in a desert. If I would write a paper claiming that a pure $SU(3)$ Yang-Mills theory could explain the strong interactions, it would be directly dismissed. I was made to believe that the real question was not the scaling, but the understanding of quark confinement. I felt that, as long as I could not argue why quarks are confined, nobody would listen to these ideas. Veltman convinced me that our work on gravity was more important, and it kept us quite busy.

As for what exactly happened in the States, I can only repeat the accounts given by others [43] [45]. D. Gross and F. Wilczek at Princeton set out to prove that no quantum field theory could ever yield a negative β-function, so that Bjorken scaling would require a radical departure from quantum field theory. H.D. Politzer at Harvard was given the problem as a student by his advisor S. Coleman. Both groups obtained an answer almost simultaneously, but when they compared what they had, there was an embarassment: Politzer had found a negative β, whereas the Princeton group had what they expected, a positive β. But they soon found their sign error, and the correct result was quickly agreed upon. Two papers were published in the same issue of Physical Review Letters [46], and these were the first published accounts of the negative beta function. Politzer suggested that his finding would render the non-Abelian perturbation expansion 'reliable'; this was perhaps a bit too optimistic, but, of course, perturbative approaches would indeed come within reach at the high-energy domain. It was realized that Bjorken scaling would still be violated by

---

8   I should have known that this view was not at all shared by everyone. Amplitudes with exchanges at high, spacelike values for q^2 are certainly physically relevant; indeed, very high spacelike q^2 are now directly experimentally accessible.

logarithms, but the general feeling was that the logarithms would be accounted for, as indeed they would.

When I heard about these papers, I was surprised, not about the result, which I had known all along, but about the stir they caused. And, finally, people began to talk about a pure $SU(3)$ theory for quarks. Well, I thought, they still have nothing, as Veltman had taught me, what really matters is the quark confinement problem. No real advances had been made yet. I set out to try to understand it from first principles. QCD had only come half-way.

The original calculations of the QCD β-functions were rather complicated, as was the calculation which I had presented in Marseille. In the meantime, however, a much more elegant and faster procedure had been invented, since it was observed that these β-functions are closely related to the counter terms in the dimensional regularization scheme [47], and then the calculation of these counter terms was streamlined by using background field methods [48]. The original β-functions used by Callan and Symanzik had a complicated mass dependence in them. This, however, was not essential. If one uses different definitions of the coupling constant, one can eliminate this mass dependence. The dimensional renormalization procedure could be associated with what was later called 'minimal subtraction': the pure subtraction of only the dimensional poles from the amplitudes. The β-functions then become mass-independent. The importance of this result is that the picture of what is going on physically is significantly simplified. One can read off the β-functions quickly from the Lagrangian of the theory. Only in the case that particles have a very *large* mass, do the expressions develop new large logarithms, which can be removed by considering the heavy particles to be 'decoupled', but then the coupling constants have to be redefined, and the β-functions change.

3   CONFINEMENT

Several key developments then helped us to understand confinement. The string-interpretation of the dual models gave a first clue: mesons behave as little stretches of strings, with quarks at their end points. This most beautifully squared with the observed linear relation between the mass-squared and the angular momentum of the heavy mesonic resonances, the so-called Regge trajectories[9]. This now suggests that the colored forces between quarks form *vortex lines* that connect color charges. How can one understand why these are formed? This is a long-distance problem, and precisely at these long distances (of a Fermi and more), the coupling strength grows so large that perturbative arguments will be meaningless.

Independently, Nielsen and P. Oleson [51] in Copenhagen, and B. Zumino [52] at CERN, discovered a new and extremely important feature in an *Abelian* Higgs theory: this model would allow for the existence of a magnetic vortex line that would have several features in common with the dual string: it is unbreakable, its mass-energy content is entirely due to the tension force, and it exhibits an internal Lorentz invariance (for boosts in the direction of the string). If quarks would be magnetic monopoles with respect to this spontaneously broken $U(1)$ force, you would get exactly the model you want. In fact, this model is nearly identical to much older models for superconductor material in which the Meissner effect is active. The vortices are just the well-known Abrikosov vortices formed by magnetic fields that succeed in penetrating into a superconductor.

But quarks are not monopoles. This theory was not QCD, and so it would not explain Bjorken scaling and the symmetric $SU(3)$ representations of the

---

9   The work of J. Chew [1], S. Frautschi [49] and V. Gribov [50] should be mentioned here.

hadrons. What happens if one replaces the Abelian $U(1)$ theory by a non-Abelian theory with Higgs mechanism? Does that also allow for Abrikosov vortices? I asked myself this question, and ran into a puzzle: there would be vortices, but they would differ from the Abelian ones in that now they can break. The resolution of the puzzle was revolutionary: the vortices break because they can create pairs of magnetic monopoles and antimonopoles. It was not known yet that magnetic monopoles can exist in fairly ordinary non-Abelian gauge models! I had now learned not to sit upon such a result for long. The paper was published in which I explained the nature of the magnetic monopoles in gauge theories [53]. Independently, this discovery was made in Moscow: A. Polyakov had constructed the same gauge field solution [54], although he did not immediately realize that the object carries a single magnetic charge.

Magnetic monopoles turned out to be a magnificent laboratory for the study of the remarkable topological features of gauge theories. When I was back at Utrecht, Peter Hasenfratz was a young postdoc there, and he asked me questions that made me realize the existence of a very peculiar feature in gauge theories with monopoles. We wrote a paper about it together [55]. Imagine a gauge theory without any fermions, but there are (bosonic) fields with gauge-isospin of 1/2. If the Higgs field carries isospin 1, we have magnetic monopoles with magnetic charge $2\pi/e$ and we have electrically charged particles with charge e. Both will be bosons with integral spin. However, if these two form a bound state, it will have orbital angular momentum that is half-integer [55]. Moreover, I can give beautiful arguments (but never published) that this bound state made of only bosons actually obeys Dirac-Fermi statistics.[10] I thought it was great to get fermions out of the blue, in a theory

---

10   Other arguments for this, such as the one published by A. Goldhaber [56], appear to be much more indirect and difficult to follow.

that was initially without fermions.

However, it looked as if the monopoles was a side-track. By itself it did not bring us closer to understanding quark confinement; yet it would later turn out to be a crucial ingredient, as I will explain shortly.

First, something else happened. K. Wilson was studying the renormalization group for non-Abelian gauge theories restricted on a space-time lattice. His interest was primarily in the domain of computer simulations. But he made a crucial observation [57]: if, for practical, technical reasons, you restrict (Euclidean[11]) space and time to form a lattice, you can perform a new perturbative expansion: the $1/g^2$ expansion. This large coupling constant expansion becomes accurate at large distance scales. Surprisingly, one can read off right away that the only states that one can then talk about are gauge-invariant (that is, colorless) states, formed by quarks that had to be connected to each other by line segments of the lattice, exactly as in the dual string theory! In other words, quark confinement is evident *and absolute* in the large coupling constant expansion!

In my recollection, this was the first indication that quark confinement is absolute, quarks can never escape from each other. It looks as if, for a theory on a lattice, when the coupling constant is varied from small to large, a *phase transition* occurs. QCD is said to be in the *confinement phase*. Yet, in the continuum theory, the interactions are generated at very tiny distance scales, where the coupling constant is weak, and then, by cumulative effects, they generate larger couplings at large distances. Does this justify a large coupling constant expansion for continuum theories? Not really. We were coming closer, but the exact nature of the confinement mechanism had not yet been

---

11   And, of course, by this time, the relevance of Euclidean space for the physical aspects
      of the theory was obvious.

explained convincingly.

In statistical physics, there is a famous exactly solvable model, called the Ising Model (first solved by L. Onsager [58]). The model has a dual symmetry that relates small coupling (high temperature) to strong coupling (low temperature). A phase transition occurs at the so-called duality point [59]. In a super conductor, the phase transition to superconductivity is of a comparable nature. In QCD, we have color electric fields and color magnetic fields. If there were a Higgs mechanism, it would be the magnetic fields that form vortex lines, much like the ones Nielsen, Olesen and Zumino had derived in a $U(1)$ theory with Higgs mechanism. In addition, as Polyakov and I had shown, there would be free magnetic monopoles. They would be at the end points of the magnetic vortices. The confinement mechanism that we wanted to understand, would require the color-electric field lines to form stringlike vortices. The color-electric quarks should be at the end points. Clearly, the relevant duality symmetry here is electric-magnetic duality.

Because of Dirac's relation [60] between the fundamental electric charge quantum e and the magnetic charge quantum g_m,

$$g_m e = 2 \pi n \hbar c,$$

where n is an integer, the interchange electric-magnetic also corresponds to a small coupling-large coupling interchange, just as in the Ising model. Therefore, to understand confinement, one would have to interchange the electrically charged particles with the magnetic monopoles in the theory. If, subsequently, a Higgs mechanism is assumed to take place among the magnetic monopoles, we obtain the dual analogue of the older model of Nielsen, Olesen and Zumino. The magnetic vortex in that model would now manifest itself as an electric vortex line, and the role of the monopoles in that model would now be played by the quarks. They will be held together absolutely and permanently by these vortices. Bingo, this is permanent quark confinement.

I had the opportunity to report this mechanism in my rapporteur's talk [61] at the Palermo Conference of 1975. S. Mandelstam had the same idea [62]. The next step would be to reformulate this mechanism in more precise terms. This was done by carefully rephrasing the gauge constraints of the non-Abelian theory. It had been noted by Gribov [63], that the usual procedure for fixing the gauge degree of freedom in an Abelian gauge theory leads to a complication when applied to the non-Abelian case: the Lorenz condition does *not* fix the gauge freedom completely; there is an ambiguity left. This ambiguity is harmless as long as one sticks to a perturbative expansion. There, we know exactly what kind of non-local effects are caused by gauge fixing: they are the Faddeev-Popov ghosts. The procedure of treating them is well-understood in perturbation expansion.

But in attempts to go beyond perturbation expansion, Gribov's ambiguity is important. It is then important to be able to distinguish physical states from ghosts. Ghosts arise as soon as the gauge fixing procedure requires the solution of differential equations; their kernels are the ghost propagators. If one wants to avoid ghosts altogether, one must use a *local, non-propagating* gauge fixing procedure. Indeed, it is possible exactly to give a gauge constraint that removes *the non-Abelian parts* of the local gauge invariance, without any ambiguity [64]. After such a gauge fixing procedure, one is left with an apparently Abelian effective theory, *with* in addition to the electrically charged particles, *magnetic monopoles*. They emerge as singularities due to the gauge fixing. Since the Abelian gauge group that survives is the Cartan subgroup of the original gauge group (the largest Abelian subgroup), we call this procedure the 'maximal

Abelian gauge fixing procedure', or 'the Abelian projection'.

In the maximal Abelian gauge, we find 'electrically' charged particles, 'magnetically' charged particles, each with finite masses depending on the dynamical features of the system, and massless $U(1)$ 'photons', that couple to the 'electric' and 'magnetic' particles in a totally symmetric fashion. Of course, these 'electric' and 'magnetic' charges are not the charges that couple to the ordinary photons. We should call them 'color-electric' and 'color-magnetic', and the $U(1)$ photons they couple to are really gluons. All one is now left with to do, is to invoke the Higgs mechanism in this color regime. If the Higgs mechanism applies to the color-electrically charged particles, then they will roam around freely, only interacting with each other at short ranges, since they exchange gluon-photons that now have become massive. The color-magnetic particles will be held together eternally by color-magnetic vortex lines. But the effective model in the maximal Abelian gauge equally well allows us to assume the Higgs mechanism to work on the color-magnetically charged particles, in which case the color-electrically charged particles are held together by color-electric vortices, being the dual strings. Whether the color-electric, or the color-magnetic Higgs mechanism takes place, or perhaps neither, depends on the relative strength of the color-electric and the color-magnetic charges. Due to asymptotic freedom, one generally expects the color-electric charges to become large when it is attempted to separate them by large distances; the color-magnetic charges will become small, with a tendency to be screened completely. If one assumes that the latter undergo Bose condensation, just as in super conductors and super fluids, one has permanent quark confinement.

The above is not a *proof* that quarks in QCD are confined, but it is a description of the mechanism; whether it actually occurs depends on dynamical features such as the ratios of various forces in the theory, and in most cases this can only be established by detailed numerical analysis.

This analysis has now been done. A daring experiment was first carried out by M. Creutz *et al*, on a computer [65]: they took a lattice in four-dimensional Euclidean space and defined the QCD field variables on this lattice. Because of the complexity of the problem, and the limited capacity of these early computers, the lattices had to be very modest in size. Nevertheless, their computer simulations showed clear evidence of a transition towards a confining phase. Nowadays, by using much larger and faster computers, as well as much improved analytical procedures, this transition can be studied in much more detail. One has also been able to impose maximal Abelian gauge conditions on the leading color field configurations that were obtained, and quite generally, the picture of Bose-condensing magnetic monopoles has been confirmed [66]. Today, it is fair to say that quark confinement is no longer seen as a deep puzzle in quantum field theory. We therefore understand why protons did not break into quarks in the ISR experiments at CERN [42]. If a color vortex ever breaks, it is because of quark-antiquark pair formation, so that hadrons are always kept color neutral. A vortex breaking without quark pair formation would require large-scale structures (*i.e.*, low energy excitations) of a kind that directly contradict the lattice simulations. Therefore, we do not expect that experimentalists will ever detect free quarks. If fractional electric charges will ever be observed, this will be highly interesting from a theoretical point of view, but these objects will have little to do with quarks.

4 THE $U(1)$ ANOMALY

Permanent quark confinement was not the only puzzle that had to be solved before the more tenacious critics would accept QCD as being a realistic theory. There had been another cause for concern. This was the so-called η-η'

problem. The problem was evident already to the earliest authors, Fritzsch, Gell-Mann and Leutwyler [38]. The QCD Lagrangian correctly reflects all known symmetries of the strong interactions:

- baryon conservation, symmetry group $U(1)$;

- isospin, symmetry group $SU(2)$, broken by electromagnetism, and also a small mass difference between the up and the down quark;

- an $SU(3)$ symmetry, broken more strongly by the strange quark mass;

- spontaneously broken chiral $SU(2) \times SU(2)$, also explicitly broken by the mass terms of the up and the down quark. The explicit chiral symmetry breaking is reflected in the small but non-vanishing value of the pion mass-squared, relative to the mass-squared of other hadrons;

- spontaneously broken chiral $SU(3) \times SU(3)$, explicitly broken down to $SU(2) \times SU(2)$ by the strange quark mass, which leads to a relatively light kaon.

Furthermore, the hadron spectrum is reasonably well explained by extending these symmetries to an approximate $U(6)$ symmetry spanned by the d, u and s quark, each of which may have their spins up or down.

The problem, however, is that the QCD Lagrangian also shows a symmetry that is known to be badly broken in the observed hadron spectrum: instead of chiral $SU(2) \times SU(2)$ symmetry (together with a single $U(1)$ for baryon number conservation), the Lagrangian has a chiral $U(2) \times U(2)$, and consequently, it suggests the presence of an additional axial $U(1)$ current that is approximately conserved. This should imply that a fourth pseudoscalar particle (the quantum numbers are those of the η particle) should exist that is as light as the three pions. Yet the η is considerably heavier: 549 MeV, instead of

135 MeV. For similar reasons, one should expect a ninth pseudoscalar boson that is as light as the kaons. The only candidate for this would be one of the heavy mesons, originally called the X^0 meson. When it was discovered that this meson can decay into two photons [67], and therefore had to be a pseudoscalar, it was renamed as η'. However, this η' was known to be as heavy as 958 MeV, and therefore it appeared to be impossible to accommodate for the large chiral $U(1)$ symmetry breaking by modifying the Lagrangian. Also, the decay ratios — especially the radiative decays — of η and η' appeared to be anomalous, in the sense that they did not obey theorems from chiral algebra [68].

A possible cure for this disease was also recognised: the Adler-Bell-Jackiw anomaly. The divergence of the axial vector current is corrected by quantum effects:

$$\partial_\mu J_\mu^A = \frac{g^2}{16\pi^2} \varepsilon_{\mu\nu\alpha\beta} \, Tr \, G_{\mu\nu} G_{\alpha\beta} \; ,$$

where J_μ^A is the axial vector current, $G_{\mu\nu}$ the Yang-Mills gluon field, and g the strong coupling constant. So, the axial current is not conserved. Then, what is the problem?

The problem was that, in turn, the r.h.s. of this anomaly equation can also be written as a divergence:

$$\varepsilon_{\mu\nu\alpha\beta} \, Tr \, G_{\mu\nu} G_{\alpha\beta} = \partial_\mu K_\mu ,$$

where K_μ is the Chern-Simons current. K_μ is not gauge-invariant, but it appeared that the latter equation would be sufficient to render the η particle as light as the pions. Why is the η so heavy?

There were other, related problems with the η and η' particles: their mixing. Whereas the direct experimental determination of the ω–ϕ mixing [69] allowed to conclude that in the octet of vector mesons, ω and ϕ mix in accordance to their quark contents:

$$\omega = \frac{1}{\sqrt{2}}\,(u\bar{u} + d\bar{d})\,, \quad \phi = s\bar{s}\,,$$

the η particle is strongly mixed with the strange quarks, and η' is nearly an $SU(3)$ singlet:

$$\eta \approx \frac{1}{\sqrt{3}}\,(u\bar{u} + d\bar{d} - s\bar{s})\,, \quad \eta' \approx \frac{1}{\sqrt{6}}\,(u\bar{u} + d\bar{d} + 2s\bar{s})\,.$$

Whence this strong mixing?

Several authors came with possible cures. Kogut and Susskind [70] suggested that the resolution came from the quark confinement mechanism, and proposed a subtle procedure involving double poles in the gluon propagator. Weinberg [71] also suggested that, somehow, the would-be Goldstone boson should be considered as a ghost, cancelling other ghosts with opposite metrics. My own attitude [61] was that, since K_μ is not gauge-invariant, it does not obey the boundary conditions required to allow one to do partial integrations, so that it was illegal to deduce the presence of a light pseudoscalar.

Just as it was the case for the confinement problem, the resolution to this $U(1)$ problem was to be found in the very special topological structure of the non-Abelian forces. In 1975, a topologically non-trivial field configuration in four-dimensional Euclidean space was described by four Russians, A.A. Belavin, A.M. Polyakov, A.S. Schwarz and Yu.S. Tyupkin [72]. It was a localized configuration that featured a fixed value for the integral

$$\int d^4x\,\varepsilon_{\mu\nu\alpha\beta}\,Tr\,G_{\mu\nu}G_{\alpha\beta} = \frac{32\pi^2}{g^2}\,.$$

The importance of this finding is that, since the solution is localized, it obeys all physically reasonable boundary conditions, and yet the integral does not vanish. Therefore, the Chern-Simons current does not vanish at infinity. Certainly, this thing had to play a role in the violation of the $U(1)$ symmetry.

This field configuration, localized in space as well as in time, was to be called "instanton" later [73]. Instantons are sinks and sources for the chiral current. This should mean that chirally charged fermions are created and destroyed by instantons. How does this mechanism work?

Earlier, R. Jackiw and C. Rebbi [74] had found that the Dirac equation for fermions near a magnetic monopole shows anomalous zero-energy solutions, which implies that magnetic monopoles can be given fractional chiral quantum numbers. We now discovered that also near an instanton, the Dirac equation shows special solutions, which are fermionic modes with vanishing action [73]. This means that the contribution of fermions to the vacuum-to-vacuum amplitude turns this amplitude to zero! Only amplitudes in which the instanton creates or destroys chiral fermions are unequal to zero. The physical interpretation of this was elaborated further by Russian investigators, by Jackiw, Nohl and Rebbi [75], and by C. Callan, R. Dashen and D. Gross [76]: instantons are tunnelling events. Gauge field configurations tunnel into other configurations connected to the previous ones by topologically non-trivial gauge transformations. During this tunnelling process, one of the energy levels produced by the Dirac equation switches the sign of its energy. Thus, chiral fermions can pop up from the Dirac sea, or disappear into it. In a properly renormalized theory, the number of states in the Dirac sea is precisely defined, and adding or subtracting one state could imply the creation or destruction of an antiparticle. This is why the original Adler-Bell-Jackiw anomaly was first found to be the result of carefully renormalizing the theory.

With these findings, effective field theories could be written down in such a way that the contributions from instantons could be taken into account as extra terms in the Lagrangian. These terms aptly produce the required mass terms for the η and the η', although it should be admitted that quantitative agreement is difficult to come by; the calculations are exceptionally complex and involve the cancellation of many large and small numerical coefficients against each other. But it is generally agreed upon (with a few exceptions) that the η and η' particles behave as they should in QCD[12].

5   REFINEMENTS

Time has come to address more detailed questions. First, we would also like to understand why chiral $SU(2) \times SU(2)$ and chiral $SU(3) \times SU(3)$ are spontaneously broken. Banks and Susskind [79] came with a neat argument. In a pseudoscalar meson, the quarks are in an S-state, which means that they move primarily in the radial direction. In the color gauge field, the quark helicity should be conserved, so when the radial velocity switches sign, so should the quark spin. Such a simultaneous switch appears to be difficult to understand. It is more likely that the quarks simply continue their ways into the Dirac sea while other inhabitants of this sea take their place on the way back. This, these authors argue, implies that $q\bar{q}$ must have a large expectation value in the vacuum. This is spontaneous chiral symmetry breaking.

A second argument came from the effective Lagrangians for the hadrons. The terms due to instantons that represent the breakdown of chiral $U(1)$ conservation, also add to the effective potentials terms that induce spontaneous symmetry breaking. But a more convincing argument comes from a careful

---

12   There were protests. R. Crewther continued to disagree [77]. It was hard to convince him [78].

consideration of all possible Adler-Bell-Jackiw anomalies that might arise in the presence of "weak spectator gauge fields" (such as ordinary electroweak fields, but one may also imagine superweak interactions that have not yet actually been detected). We simply require that the effective theories reproduce the correct anomalies. This is called the anomaly matching condition [80]. It turns out to be extremely restrictive. Although it does allow chiral $SU(2) \times SU(2)$ to be in the Wigner mode, this is not the case [13] for chiral $SU(3) \times SU(3)$. If we assume that the small value of the strange quark mass has little effect on the vacuum expectation values of non-strange operators, chiral symmetry must be spontaneously broken.

Can we analyze QCD with more numerical precision, and formulate new predictions? Enormous amounts of work have been done. The simulations of QCD on a lattice are making good progress. On the one hand, computers are getting larger and faster. This helps, but not in a very spectacular way. The simulation programs must be improved, and this is also being done. The trick is to suppress two sources of error: one is the effects of the finiteness of the lattice size. So-called improved actions [81] are being designed that are being chosen in such a manner that the small-lattice-size limit converges as fast as is possible. Secondly, the effects due to the small volume of the total system should be suppressed: usually the size is chosen in such a way that the particle under consideration, say the proton, just barely fits in there. But we can try to predict the effects of the boundary, so that they can be corrected for. This, however, is usually considered to be very difficult.

Various different models have been proposed that are related to QCD proper, but have been modified in such a way that analytical calculations are possible. A first simplification is to replace the number N of color degrees of freedom, which is 3 for the physical theory, by a number tending to infinity. In the $N \to \infty$ limit, mesons become non-interacting [82], and baryons behave as solitons (a proposal to treat baryons as solitons in a meson theory has been proposed long ago by T.H. Skyrme [83], and so these baryons are now called skyrmions). The coupling constant between the mesons is essentially $1/N$. Thus, it was proposed to consider the $1/N$ expansion. Unfortunately, the $N \to \infty$ limit is still an extremely complicated theory, and closed expressions for, say, the meson mass spectrum have not been found. It is tempting to suspect that some modified string theory should accommodate for them, but this is not known, in spite of numerous attempts and conjectures.

What can be solved exactly is the $N \to \infty$ limit in two dimensions (one space, one time dimension) [84]. It is comforting to observe that in this limit one not only sees confinement, with a calculable spectrum of linearly rising trajectories of mesonic resonances, but also chiral symmetry is seen to be broken spontaneously, with the pion mass-squared being linearly proportional to the up and the down quark masses, when these are small. Working out the Skyrmion spectrum in this model is much more difficult.

Recently, considerable progress was reported concerning supersymmetric versions of QCD [85]. In many supersymmetric models, the vacuum state can be characterized accurately, and many of our insights concerning the confinement mechanism, the role of instantons and chiral symmetry breaking could be verified explicitly [86].

6   COMPARING WITH EXPERIMENT

The experimental data that were used for the construction of QCD were the spectra of the lowest hadronic states, and the linearity of the so-called Regge trajectories (*i.e.* the linear relation between mass-squared and angular momentum for all mesonic and baryonic resonances). QCD, having only one coupling parameter that fixes the scale of the interactions, and for the rest requiring only a few quark masses, explains these features remarkably well. But the truly new predictions are the perturbative expressions that should gain more and more accuracy at higher energies. The first spectacular event was the J/ψ discovery. The exotic features of J/ψ were not predicted[13] but they should have been. Due to their large masses, the charmed quark and antiquark move in a region that is much smaller than that of all other mesons. Hence they probe QCD where its coupling is relatively small. The decay rate, of which the first stage leads to an intermediate state with three gluons,

$$c\bar{c} \to g\,g\,g$$

is governed by a factor g^6, where g is the QCD coupling constant. Furthermore, the $c\bar{c}$ wave function tends to spread, yielding even more powers of g. So, if g decreases, the decay rate drops very fast. This explains the small decay rate of J/ψ. But, since we can here compute the interquark potential and the wave functions, many of the details of the other bound states could be predicted. The story was repeated for the beauty quark, and, much more recently, for the top quark, for which the gauge coupling is even further reduced.

But it should be possible to check QCD more directly at very high energies. When quarks or leptons hit each other at small impact parameters, we get events where quarks and gluons are produced with large transverse momenta. A theory was developed that describes how these quarks and gluons evolve and turn into hadronic states. They form showers of hadrons, called jets

---

13   They nearly were, by Th. Appelquist and H.D. Politzer [87], but their paper about the $c\bar{c}$ bound state came after J/ψ had been discovered.

[88]. The first phases of this process can be handled by perturbation expansion, whereas the final phases must be universal for all types of jets. This way, the properties of the jets initiated by quarks and gluons could be predicted. If, during the initial phase, additional gluons are produced by radiative effects, multi-jet events will result. Apparently, however, theoreticians were unable to prescribe what experimentalists had to look for to establish the universal nature of these final interactions. The experimental results were discouraging; scattering experiments yielded different final states for each pair of interacting particles. So it happened that these aspects of QCD had to wait until experimentalists themselves came with the right idea [89]. The showers come with what is now called an "effective energy", and, in terms of this quantity, universality could be established [90].

On a statistical basis, one can now distinguish quark jets from gluon jets, and this enables us to check QCD predictions at high energies. Hadronic decay events of the Z and W bosons for instance, will be good laboratories, since in these cases the initial states are very precisely known. If the initial particles are nucleons, one has to insert information about the nucleonic wave functions in terms of quarks and gluons. The ratios of the occurence of sea quarks and valence quarks have been studied, but other aspects of the nucleonic wave function need further clarification.

When at very high energies, larger nuclei are collided one against the other, one expects an effective heating of hadronic matter. At sufficiently high temperatures, transitions may take place in which quarks get deconfined. Experimental signals for such a new plasmalike phase have been investigated. The importance of investigating the quark-gluon plasma lies in the fact that during the evolution of the early universe, matter may have been in this phase for a short time.

7   THE θ ANGLE

In the absence of massless fermions, the color fields may freely perform transitions from one gauge configuration into a topologically rotated one. Since these gauge rotations are an exact invariance of the system, forming the (additive) group of integers, $\mathbf{Z}$, the vacuum state must be a representation of this gauge group, which is characterized by an angle θ. If there are massless fermions, the effects of this angle disappear [73] [75] [76], while the angle itself merely fixes the direction in which chiral symmetry is explicitly broken. One can also say that, with respect to the axis specified by the θ angle, a fermionic mass term becomes a complex number:

$$ m \rightarrow m \ (\cos \theta + i\gamma^5 \sin \theta). $$

The imaginary part of this mass violates CP invariance, and so, CP invariance breaking is predicted by QCD, but the amount of breaking — being proportional to sin θ — is unpredictable. Since explicit CP violation also occurs elsewhere in particle physics (notably the kaon system), the choice $\theta = 0$ is not an option. Virtual kaons will renormalize θ into a value different from zero.

One of our problems is to determine θ from experimental observations. Strong CP breaking, however, has not (yet) been observed. This implies that θ must be very small.

The smallness of θ is puzzling. Is there a conspiracy that tunes θ towards zero? A popular proposal is that there is a spontaneous symmetry breaking effect acting on a new pseudoscalar field that is coupled to θ, the so-called axion. The axion may be a relatively light, not strongly interacting pseudoscalar elementary particle. It has been searched for for a long time, without success.

The θ parameter cannot be tuned by an experimentalist; it is a God-given number. But if we could tune it, it would be interesting to calculate what might happen as θ varies. Theoretical computations have been made. Notably, one predicts one or more phase transitions as θ runs from zero to 2π. If nowhere else, then a phase transition must occur [64] [91] at $\theta = \pi$. At this point, the monopole field that causes confinement by undergoing the Higgs mechanism, must switch places with a monopole-gluon bound state.

θ-dependence plays an important role in the supersymmetric models, where $\dfrac{1}{8\pi^2 g^2} + \theta i$ is taken to be the primary physical parameter.

Studying these θ effects will also be of importance because a repetition of the story is expected for the electroweak theory. Here, instanton-related effects will be excessively weak, except perhaps at very high energies and very high multiplicities of the final states. In this theory, instantons induce a symmetry breaking that is more spectacular than in the strong interaction theory: it is the baryon number itself that is destroyed. Still earlier in the evolution of the universe, when quarks and fermions, together with photons and intermediate vector bosons, formed an electroweak plasma [92], instantons were creating and destroying baryonic matter, by forming metastable intermediate solitonic particles, called 'sphalerons' [93]. The observed baryon excess in the present universe must be due to an asymmetry in these instanton related effects. This asymmetry can only be explained if we assume that the universe at this state was far away from equilibrium. Baryons probably were created in excess of antibaryons during the time that bubbles of true vacuum configurations appeared in a supercooled but symmetric pseudovacuum. The required asymmetry occurred in the rapidly expanding bubble walls [94].

8   REMAINING QUESTIONS

QCD is an extremely complex theory, in spite of the extreme simplicity of its basic equations. Both from the experimental side, and from the theoretical side, important issues have to be resolved. Some theoretical questions are extremely fundamental. To start with, it has not been proven that QCD is mathematically airtight. We would like to have prescriptions for computing hadronic states, or wave functions, for ordinary particles such as the pion and the proton, as well as scattering matrix amplitudes, which should be infinitely accurate, *at least in principle*. This is a hard question, considering the fact that even for the classical problem of N gravitating particles such questions are not easy to answer. Next, we would like to possess decent approximation methods that convert reasonably rapidly to the desired numerical answers. Presently, we do have the Monte Carlo schemes on a lattice, but here convergence is painfully slow.

It is a big challenge to analyse in more detail the large N expansion (N being the number of colors) [82]. As stated before, in the large N limit, mesons are non-interacting, so it should be relatively easy to compute their spectrum. Unfortunately, this is not easy. One expects that freely moving strings, modified by stiffness terms in their Lagrangian, should describe these states, but this has not been shown. It is also worth-while to continue efforts to first put the $N \rightarrow \infty$ domain of QCD on a rigorous mathematical footing [95]. That would leave us with a $1/N^2$ expansion, which, with $N = 3$, may lead to results that can be checked experimentally.

Extremely difficult from the theoretical point of view, is the treatment of the many particle in, many particle out, scattering process. Even if the initial state is simple, like an electron-positron annihilation event, the many-particle final state is difficult to compute, or even estimate. Notably the θ-dependent effects are difficult. This was dramatically exhibited in the electroweak theory, where, in spite of the smallness of the coupling constants, transitions into multiples of $1/\alpha$ final particles, generated hot disputes as to whether the instanton effects will be negligible, or perhaps even dominant. The quark-gluon plasma is an extreme example of a multi-particle problem. It will have to be handled by statistical quantum field theory, a doctrine that has already matured.

Experimentalists should study the complete arena of ultra-high multiplicity events, in particular when a simple initial state, such as e^+e^-, gives rise to a very high multiplicity of low energy hadronic particles.

Lattice theorists should continue to produce faster algorithms, devise improved actions, and make models for the boundary effects of the (periodic) lattice walls, which hopefully will enable them to perform precision tests and comparisons with experiment.

9   CONCLUSION

Quantum Chromodynamics is a magnificent theory. Not one physicist, or even one group of physicists, can claim the discovery of this theory. Many new insights were needed to convince the scientific community that this set of equations, a pure $SU(3)$ Yang-Mills theory with a handful of massive fermions and essentially no scalars, can completely account for all observed strong interactions. It is now one of the best established branches of the Standard Model, but it may still keep theoreticians as well as experimentalists busy for quite some time to come.

REFERENCES

[1]   G. Chew, in *The Analytic S-Matrix*, W.A. Benjamin Inc. (1966).

[2]   M. Goldberger, in *The Quantum Theory of Fields*, Proceedings of the 12th Solvay Conference, Interscience, New York (1961).

[3]   M. Gell-Mann, "The Eightfold Way — A Theory of Strong-Interaction Symmetry", California Institute Technology Synchrotron Lab. Report $\underline{20}$ (1961); Y. Ne'eman, *Nuclear Physics* $\underline{26}$ (1961) 222; see also: M. Gell-Mann and Y. Ne'eman, "The Eightfold Way", Benjamin, New York (1964).

[4]   M. Gell-Mann, *Physics Letters* $\underline{8}$ (1964) 214; G. Zweig, Erice Lecture 1964, in *Symmetries in Elementary Particle Physics*, A. Zichichi, Ed., Academic Press, New York, London, 1965, and CERN Report No. TH 401, 4R12 (1964), unpublished.

[5]   M. Gell-Mann, *Physics* $\underline{1}$ (1964) 63; R. Dashen and M. Gell-Mann, *Physical Review Letters* $\underline{17}$ (1966) 340; S. Adler, *Physical Review* $\underline{140B}$ (1965) 736; W.I. Weisberger, *Physical Review* $\underline{143B}$ (1966) 1302.

[6]   R.P. Feynman, in *The Quantum Theory of Fields*, Proceedings of the 12th Solvay Conference, Interscience, New York (1961). See also: T.Y. Cao and S.S. Schweber, *The Conceptual Foundations and Philosophical Aspects of Renormalization Theory, Synthese* $\underline{97}$ (1993) 33, Kluwer Academic Publishers, The Netherlands.

[7]   V. Alessandrini, D. Amati, M. Le Bellac and D. Olive, *Physics Reports* $\underline{1C}$ (1971) 269; J.H. Schwarz, *Physics Reports* $\underline{8C}$ (1973) 270; S. Mandelstam, *Physics Reports* $\underline{13C}$ (1974) 259.

[8]   G. Veneziano, *Nuovo Cimento* $\underline{57A}$ (1968) 190.

[9]   Z. Koba and H.B. Nielsen, *Nuclear Physics* $\underline{B10}$ (1969) 633, *ibid.* $\underline{B12}$ (1969) 517; $\underline{B17}$ (1970) 206; *Zeitschrift für Physik* $\underline{229}$ (1969) 243.

[10]   H.B. Nielsen, "An Almost Physical Interpretation of the Integrand of the n-point Veneziano model", *XV Int. Conf. on High Energy Physics*, Kiev, USSR, 1970; Nordita Report (1969), unpublished; See also: D.B. Fairlie and H.B. Nielsen, *Nuclear Physics* $\underline{B20}$ (1970) 637.

[11]   H.B. Nielsen and L. Susskind, CERN preprint TH 1230 (1970), Y. Nambu, Proceedings International Conference on *Symmetries and Quark models*, Wayne State University (1969); Lectures at the Copenhagen Summer Symposium (1970); T. Goto, *Progr. Theor. Phys.* 46 (1971) 1560; L. Susskind, *Nuovo Cimento* 69A (1970) 457; *Physical Review* 1 (1970) 1182.

[12]   M. Gell-Mann and M. Lévy, *Nuovo Cimento* 16 (1960) 705.

[13]   B.W. Lee, "Chiral Dynamics", Gordon and Breach, New York (1972).

[14]   J.-L. Gervais and B.W. Lee, *Nuclear Physics* B12 (1969) 627; J.-L. Gervais, Cargèse lectures, July 1970.

[15]   K. Symanzik, Cargèse lectures, July 1970.

[16]   G. 't Hooft, "The Limits of our Imagination in Elementary Particle Theory", to be published in *35th course, Highlights: 50 Years Later*, Erice, Aug.-Sept. 1997.

[17]   J.D. Bjorken, *Physical Review* 179 (1969) 1547.

[18]   R.P. Feynman, *Physical Review Letters* 23 (1969) 337.

[19]   C.G. Callan, *Physical Review* D2 (1970) 1541.

[20]   K. Symanzik, *Commun. Math. Phys.* 16 (1970) 48; *ibid.* 18 (1970) 227, *ibid.* 23 (1971) 49.

[21]   M. Gell-Mann and F. Low, *Physical Review* 95 (1954) 1300.

[22]   E.C.G. Stueckelberg and A. Peterman, *Helv. Phys. Acta* 26 (1953) 499; N.N. Bogoliubov and D.V. Shirkov, *Introduction to the theory of quantized fields*, Interscience, New York (1959); A. Peterman, *Physics Reports* 53C (1979) 157.

[23]   S.L. Glashow, *Nuclear Physics* 22 (1961) 579.

[24]   A. Salam and J.C. Ward, *Physics Letters* 13 (1964) 168, A. Salam, Nobel Symposium 1968, ed. N. Svartholm.

[25]   S. Weinberg, *Physical Review Letters* 19 (1967) 1264.

[26]   R.P. Feynman, *Acta Phys. Polonica* 24 (1963) 697.

[27] B.S. DeWitt, *Physical Review Letters* 12 (1964) 742, *id.*, *Physical Review* 160 (1967) 1113; *ibid.* 162 (1967) 1195, 1239.

[28] L.D. Faddeev and V.N. Popov, *Physical Review* 25B (1967) 29; L.D. Faddeev, *Theor. and Math. Phys.* 1 (1969) 3 (in Russian), *Theor. and Math. Phys.* 1 (1969) 1 (Engl. transl.).

[29] E.S. Fradkin and I.V. Tyutin, *Physical Review* D2 (1970) 2841.

[30] M. Veltman, *Physica* 29 (1963) 186, *Nuclear Physics* B7 (1968) 637; J. Reiff and M. Veltman, *Nuclear Physics* B13 (1969) 545; M. Veltman, *Nuclear Physics* B21 (1970) 288; H. van Dam and M. Veltman, *Nuclear Physics* B22 (1970) 397.

[31] G. 't Hooft, *Nuclear Physics* B33 (1971) 173.

[32] A. Slavnov, *Theor. Math. Phys.* 10 (1972) 153 (in Russian), *Theor. Math. Phys.* 10 (1972) 99 (Engl. Transl.); J.C. Taylor, *Nuclear Physics* B33 (1971) 436.

[33] C. Becchi, A. Rouet and R. Stora, *Commun. Math. Phys.* 42 (1975) 127; *id.*, *Annals of Physics* (N.Y.) 98 (1976) 287; I.V. Tyutin, *Lebedev Prepr.* FIAN 39 (1975), unpublished.

[34] G. 't Hooft, *Nuclear Physics* B35 (1971) 167.

[35] G. 't Hooft and M. Veltman, *Nuclear Physics* B44 (1972) 189; C.G. Bollini and J.J. Giambiagi, *Physics Letters* 40B (1972) 566; J.F. Ashmore, *Lettere al Nuovo Cimento* 4 (1972) 289.

[36] T. Massam and A. Zichichi, *Nuovo Cimento* 43 (1966) 1137, and *Lettere al Nuovo Cimento* 1 (1969) 387.

[37] *See also*: H.J. Lipkin, in *Physique Nucléaire*, Les-Houches 1968, ed. C. DeWitt and V. Gillet, Gordon and Breach, N.Y. (1969) 585; H.J. Lipkin, *Physics Letters* 45B (1973) 267; Y. Nambu, in *Preludes in Theoretical Physics*, ed. A. de-Shalit et al, North Holland Pub. Comp., Amsterdam (1966) 133.

[38] H. Fritzsch, M. Gell-Mann and H. Leutwyler, *Physics Letters* 47B (1973) 365. See also: H. Lipkin, *Physics Letters* 45B (1973) 267.

[39]   O.W. Greenberg, *Physical Review Letters* 13 (1964) 598.

[40]   M.Y. Han and Y. Nambu, *Physical Review* 139 B (1965) 1006.

[41]   G. 't Hooft, *Acta Phys. Austr., Suppl.* 22 (1980) 531.

[42]   T. Massam and A. Zichichi, CERN Report, ISR Users Meeting 10-11 June 1968 (unpublished); M. Basile, G. Cara Romeo, L. Cifarelli, P. Giusti, T. Massam, F. Palmonari, G. Valenti and A. Zichichi, *Nuovo Cimento* 40A (1977) 41; the complete set of references is reported in this volume.

[43]   See: D.J. Gross, in *The Rise of the Standard Model*, Cambridge University Press (1997) 199; S. Coleman and D.J. Gross, *Physical Review Letters* 31 (1973) 851.

[44]   K. Symanzik, in Proceedings Marseille Conference 19-23 June 1972, ed. C.P. Korthals Altes; *id.*, *Lettere al Nuovo Cimento* 6 (1973) 77.

[45]   S. Coleman, private communication.

[46]   D.J. Gross and F. Wilczek, *Physical Review Letters* 30 (1973) 1343; H.D. Politzer, *Physical Review Letters* 30 (1973) 1346.

[47]   G. 't Hooft, *Nuclear Physics* B61 (1973) 455.

[48]   G. 't Hooft, *Nuclear Physics* B62 (1973) 444.

[49]   S.C. Frautschi, *Regge Poles and S-Matrix Theory*, W.A. Benjamin Inc. (1963).

[50]   V. Gribov, *ZhETF* 41 (1961) 1962 [*SJETP* 14 (1962) 1395]; V.N. Gribov and D.V. Volkov, *ZhETF* 44 (1963) 1068.

[51]   H.B. Nielsen and P. Olesen, *Nuclear Physics* B61 (1973) 45.

[52]   B. Zumino, in *Renormalization and Invariance in Quantum Field Theory*, NATO Adv. Study Institute, Capri, 1973, Ed. R. Caianiello, Plenum (1974) 367.

[53]   G. 't Hooft, *Nuclear Physics* B79 (1974) 276.

[54]   A.M. Polyakov, *JETP Lett.* 20 (1974) 194.

[55]   P. Hasenfratz and G. 't Hooft, *Physical Review Letters* 36 (1976) 1119.

[56] A.Goldhaber, *Physical Review Letters* <u>36</u> (1976) 1122.

[57] K.G. Wilson, *Physical Review* <u>D10</u> (1974) 2445.

[58] L. Onsager, *Physical Review* <u>65</u> (1944) 117; B. Kaufman, *Physical Review* <u>76</u> (1949) 1232; B. Kaufman and L. Onsager, *Physical Review* <u>76</u> (1949) 1244.

[59] M.A. Kramers and G.H. Wannier, *Physical Review* <u>60</u> (1941) 252, 263; L. Kadanoff, *Nuovo Cimento* <u>44B</u> (1966) 276.

[60] P.A.M. Dirac, *Proc. Roy. Soc.* <u>A133</u> (1934) 60; *Physical Review* <u>74</u> (1948) 817.

[61] G. 't Hooft, "Gauge Theories with Unified Weak, Electromagnetic and Strong Interactions", in *E.P.S. Int. Conf. on High Energy Physics*, Palermo, 23-28 June 1975, Editrice Compositori, Bologna 1976, A. Zichichi Ed.

[62] S. Mandelstam, *Physics Letters* <u>B53</u> (1975) 476; *Physics Reports* <u>23</u> (1978) 245.

[63] V. Gribov, *Nuclear Physics* <u>B139</u> (1978) 1.

[64] G. 't Hooft, *Nuclear Physics* <u>B190</u> (1981) 455.

[65] M. Creutz, L. Jacobs and C. Rebbi, *Physical Review Letters* <u>42</u> (1979) 1390.

[66] A.J. van der Sijs, invited talk at the Int. RCNP Workshop on Color Confinements and Hadrons - Confinement 95, March 1995, RCNP Osaka, Japan (hep-th/9505019).

[67] D. Bollini, A. Buhler-Broglin, P. Dalpiaz, T. Massam, F. Navach, F.L. Navarria, M.A. Schneegans and A. Zichichi, *Nuovo Cimento* <u>58A</u> (1968) 289; the complete set of references is reported in this volume.

[68] A. Zichichi, *Proceedings of the 16th International Conference on "High Energy Physics"*, Batavia, IL, USA, 6-13 Sept. 1972 (NAL, Batavia, 1973), Vol <u>1</u>, 145.

[69] D. Bollini, A. Buhler-Broglin, P. Dalpiaz, T. Massam, F. Navach, F.L. Navarria, M.A. Schneegans and A. Zichichi, *Nuovo Cimento* <u>57A</u> (1968) 404.

[70] J. Kogut and L. Susskind, *Physical Review* D9 (1974) 3501, *ibid.* D10 (1974) 3468, *ibid.* D11 (1975) 3594.

[71] S. Weinberg, *Physical Review* D11 (1975) 3583.

[72] A.A. Belavin, A.M. Polyakov, A.S. Schwartz and Y.S. Tyupkin, *Physics Letters* 59 (1975) 85.

[73] G. 't Hooft, *Physical Review Letters* 37 (1976) 8; *Physical Review* D14 (1976) 3432; Err. *Physical Review* D18 (1978) 2199.

[74] R. Jackiw and C. Rebbi, *Physical Review* D13 (1976) 3398.

[75] R. Jackiw, C. Nohl and C. Rebbi, various Workshops on QCD and solitons, June-Sept. 1977; R. Jackiw and C. Rebbi, *Physical Review Letters* 37 (1976) 172.

[76] C.G. Callan, R. Dashen and D. Gross, *Physics Letters* 63B (1976) 334.

[77] R. Crewther, *Physics Letters* 70B (1977) 359; *Rivista del Nuovo Cimento* 2 (1979) 63; R. Crewther, in *Facts and Prospects of Gauge Theories*, Schladming 1978, ed. P. Urban (Springer-Verlag 1978); *Acta Phys. Austriaca* Suppl. XIX (1978) 47.

[78] G. 't Hooft, *Physics Reports* 142 (1986) 357.

[79] L. Susskind, private communication.

[80] G. 't Hooft, in *Recent Developments in Gauge Theories*, Cargèse 1979, ed. G. 't Hooft et al., Plenum Press, New York (1980), Lecture III, "Naturalness, chiral symmetry and spontaneous chiral symmetry breaking", reprinted in: *Dynamical Symmetry Breaking, a Collection of reprints*, ed. A. Fahri et al., World Scientific, Singapore, Cambridge, (1982) 345.

[81] K. Symanzik, *Nuclear Physics* B226 (1983) 187, 205; M. Lüscher and P. Weisz, *Nuclear Physics* B266 (1986) 309; G.P. Lepage and P.B. Mackenzie, *Physical Review* D48 (1993) 2250, and many more recent publications, see J. Snippe, *The Uses of Improved Actions in Lattice Gauge Theory*, Leiden thesis, February 19, 1997.

[82] G. 't Hooft, *Nuclear Physics* B72 (1974) 461.

[83] T.H. Skyrme, *Proc. R. Soc.* A260 (1961) 127.

[84] G. 't Hooft, *Nuclear Physics* B75 (1974) 461.

[85] N. Seiberg, *Physics Letters* 318B (1993) 469; N. Seiberg and E. Witten, RU-94-52, IAS-94-43, hep-th/9407087.

[86] M. Shifman, in *Confinement, Duality and Non perturbative Aspects of QCD*, P. van Baal ed., Cambridge UK Workshop, June 1997, NATO ASI Series, Plenum Press, New York and London (1998) 477.

[87] Th. Appelquist and H.D. Politzer, *Physical Review Letters* 34 (1975) 43; Th. Appelquist, A. De Rújula and H.D. Politzer, *Physical Review Letters* 34 (1975) 365.

[88] G. Sterman and S. Weinberg, *Physical Review Letters* 39 (1977) 1436.

[89] G. Veneziano, Report in this volume.

[90] M. Basile, G. Cara Romeo, L. Cifarelli, A. Contin, G. D'Alì, P. Di Cesare, B. Esposito, P. Giusti, T. Massam, F. Palmonari, G. Sartorelli, G. Valenti and A. Zichichi, *Physics Letters* 92B (1980) 367; the complete list of references on the effective energy and universality features is reported in this volume; see also Ref. [89] and the original papers reproduced after Ref. [89].

[91] G. 't Hooft, "The confinement phenomenon in quantum field theory", in *1981 Cargèse Summer School lecture notes on fundamental interactions*, eds M. Lévy and J.-L. Basdevant, *NATO Adv. Study Inst. Series B: Physics*, vol. 85, p. 639.

[92] V.A. Kuzmin, V.A. Rubakov and M.E. Shaposhnikov, *Physics Letters* B155 (1985) 36.

[93] F.R. Klinkhamer and N.S. Manton, *Physical Review* D30 (1984) 2212.

[94] G.R. Farrar and M.E. Shaposhnikov, *Physics Letters* 70 (1993) 2833.

[95] G. 't Hooft, *Commun. Math. Phys.* 86 (1982) 449; *ibid.* 88 (1983) 1; G. 't Hooft, "Planar diagram field theories", in *Progress in Gauge Field Theory, NATO Adv. Study Inst. Series*, eds. G. 't Hooft et al, Plenum (1984) 271.

*DETAILED REFERENCES OF THE PAPERS
QUOTED BY G. 't HOOFT CONCERNING THE
EXPERIMENTAL CONTRIBUTIONS TO THE QUARK
CONFINEMENT AND TO THE U(1) ANOMALY
BY A. ZICHICHI AND COLLABORATORS*

DETAILED REFERENCES OF THE PAPERS QUOTED BY G. 't HOOFT
CONCERNING THE EXPERIMENTAL CONTRIBUTIONS TO
THE QUARK CONFINEMENT AND TO THE U(1) ANOMALY
BY A. ZICHICHI AND COLLABORATORS

$$\boxed{QUARK\ CONFINEMENT}$$

PRE-SLAC DISCOVERY OF SCALING

[1] HUNTING CHARGES $\pm\left(\frac{4}{3}\right)e$ IN THE COSMIC RADIATION
A. Buhler-Broglin, G. Fortunato, T. Massam and A. Zichichi
Nuovo Cimento <u>49A</u>, 209 (1967).

[2] A STUDY OF THE FRACTIONAL CHARGE CONTENT OF THE COSMIC RADIATION
A. Buhler-Broglin, P. Dalpiaz, T. Massam and A. Zichichi
Nuovo Cimento <u>51A</u>, 837 (1967).

[3] A SUMMARY OF ALL QUARKS EXPERIMENTS
A. Zichichi
Proceedings of the International Conference on "Elementary Particles", Heidelberg, Germany, 20-27 September 1967 (North-Holland, Amsterdam, 1968), 268.

THE NEW SERIES OF EXPERIMENTS
AFTER THE SLAC DISCOVERY OF SCALING

[4] QUARK SEARCH AT THE ISR
T. Massam and A. Zichichi
Presented at the ISR Users' Meeting, CERN, Geneva, Switzerland, 10-11 June 1968.

[5]   IS THE ABSENCE OF QUARKS AS WELL ESTABLISHED AS IS
      GENERALLY BELIEVED?
      M. Basile, G. Cara Romeo, L. Cifarelli, A. Contin, G. D'Alì, P. Giusti,
      T. Massam, F. Palmonari, F. Rohrbach, G. Sartorelli, G. Valenti and
      A. Zichichi
      Proceedings of the EPS European Conference on "Particle Physics", Budapest,
      Hungary, 4-9 July 1977 (KFKI, Budapest, 1977), 335.

[6]   A CRITICAL ANALYSIS OF THE QUARK STATUS
      M. Basile, G. Cara Romeo, L. Cifarelli, P. Giusti, T. Massam, F. Palmonari,
      G. Valenti and A. Zichichi
      Lettere al Nuovo Cimento 18, 529 (1977).

[7]   SEARCH FOR FRACTIONALLY CHARGED PARTICLES PRODUCED IN
      PROTON-PROTON COLLISIONS AT THE HIGHEST ISR ENERGY
      M. Basile, G. Cara Romeo, L. Cifarelli, P. Giusti, T. Massam, F. Palmonari,
      G. Valenti and A. Zichichi
      Nuovo Cimento 40A, 41 (1977).

[8]   SEARCH FOR QUARKS IN PROTON-PROTON INTERACTIONS AT
      $\sqrt{s} = 52.5$ GeV
      M. Basile, G. Cara Romeo, L. Cifarelli, A. Contin, G. D'Alì, P. Giusti,
      T. Massam, F. Palmonari, G. Sartorelli, G. Valenti and A. Zichichi
      Nuovo Cimento 45A, 171 (1978).

[9]   A SEARCH FOR QUARKS IN THE CERN SPS NEUTRINO BEAM
      M. Basile, G. Cara Romeo, L. Cifarelli, A. Contin, G. D'Alì, P. Giusti,
      T. Massam, F. Palmonari, G. Sartorelli, G. Valenti and A. Zichichi
      Nuovo Cimento 45A, 281 (1978).

THE U(1) ANOMALY

[10] EVIDENCE FOR A NEW DECAY MODE OF THE X^0-MESON: $X^0 \rightarrow 2\gamma$
D. Bollini, A. Buhler-Broglin, P. Dalpiaz, T. Massam, F. Navach,
F.L. Navarria, M.A. Schneegans and A. Zichichi
Nuovo Cimento 58A, 289 (1968).

[11] A NEW LARGE-ACCEPTANCE AND HIGH-EFFICIENCY NEUTRON
DETECTOR FOR MISSING-MASS STUDIES
D. Bollini, A. Buhler-Broglin, P. Dalpiaz, T. Massam, F. Navach,
F.L. Navarria, M.A. Schneegans, F. Zetti and A. Zichichi
Nuovo Cimento 61A, 125 (1969).

[12] UN DETECTEUR DE NEUTRONS POUR LA SPECTROMETRIE DE
MASSES MANQUANTES
D. Bollini, A. Buhler-Broglin, P. Dalpiaz, T. Massam, F. Navach,
F.L. Navarria, M.A. Schneegans et A. Zichichi
Revue de Physique Appliquée 4, 301 (1969).

[13] STUDY OF ELECTROMAGNETIC DECAYS AT CERN
D. Bollini, P. Dalpiaz, T. Massam, F. Navach, F.L. Navarria, M.A. Schneegans
and A. Zichichi
Proceedings of the International Seminar on "Vector Mesons and
Electromagnetic Interactions", Dubna, USSR, 23-26 September 1969 (JINR,
Dubna, 1969), 387.

[14] RADIATIVE DECAY OF MESONS
A. Zichichi
Proceedings of the 16th International Conference on "High-Energy Physics",
Batavia, IL, USA, 6-13 September 1972 (NAL, Batavia, 1973), Vol. 1, 145.

[15] THE DECAY MODE $\omega \rightarrow e^+ e^-$ AND A DIRECT DETERMINATION OF
THE $\omega - \phi$ MIXING ANGLE
D. Bollini, A. Buhler-Broglin, P. Dalpiaz, T. Massam, F. Navach,
F.L. Navarria, M.A. Schneegans and A. Zichichi
Nuovo Cimento <u>57A</u>, 404 (1968).

[16] AN APPARATUS OF THE NBC TYPE AND THE PHYSICS RESULTS
OBTAINED
A. Zichichi
Annals of Physics <u>66</u>, 405 (1971).

[17] A REVIEW OF THE RESULTS OBTAINED BY THE BOLOGNA-CERN
MULTIPURPOSE NBC SET-UP
M. Basile, D. Bollini, A. Buhler-Broglin, P. Dalpiaz, P.L. Frabetti, T. Massam,
F. Navach, F.L. Navarria, M.A. Schneegans, F. Zetti and A. Zichichi
Proceedings of the 3rd International Conference on "Experimental Meson
Spectroscopy", Philadelphia, PA, USA, 28-29 April 1972 (AIP, New York,
1972), 147.

THE EFFECTIVE ENERGY AND THE
UNIVERSALITY FEATURES IN QCD PROCESSES

Gabriele Veneziano

CERN, Geneva, Switzerland

THE EFFECTIVE ENERGY AND THE
UNIVERSALITY FEATURES IN QCD PROCESSES

Gabriele Veneziano
CERN, Geneva, Switzerland

PREAMBLE

This contribution should have been written by Professor V.N. Gribov, but his sudden, untimely death prevented him from doing so. I then consented to step in and help by collecting some of Gribov's most important remarks on the subject of these collected papers, and by adding a few thoughts of my own.

*   *   *

At first sight, there is no universality in the final-state hadrons produced in various high energy collisions at a given centre-of-mass energy E: it does make a difference whether the reaction involves, in the initial state, just leptons (e.g. e^+e^- collisions), one lepton and one hadron (e.g. ep, μp, νp collisions), or two hadrons (pp, πp, etc.). For a long time this difference (for instance in the multiplicity of produced charged hadrons) was attributed to the nature of the interaction: pp, νp, e^+e^- collisions were considered as examples of "strong", "weak" and "electromagnetic" interactions, respectively, hence very different from one another.

Moreover, different features are observed even when both initial particles are hadronic: for example, the average charged multiplicities produced when a π−meson interacts with a proton are different from those observed when an antiproton interacts with the <u>same</u> proton at the <u>same</u> E. The differences are not limited to charged particle multiplicities but extend to other properties, such as the longitudinal and transverse momentum distributions, the ratio of "charged" to "total" energy, the "planarity" of the hadronic final state, and so on.

The static properties of hadrons (masses, lifetimes, magnetic moments, etc.) could be understood thanks to theoretical developments that started with Gell-Mann and Ne'eman's flavour symmetry, $SU(3)_F$. The octet and decuplet structure of mesonic and baryonic states could then be interpreted in terms of more elementary constituents, the quarks, while the statistics problem (see below) was solved by the introduction of a global colour quantum number at first, of gluons later, and, finally, of the $SU(3)_C$ gauge theory known as Quantum Chromo-Dynamics (QCD).

In spite of this amazing progress, the dynamical properties of hadrons remained puzzling. Notwithstanding the fact that each pair of colliding particles (πp, Kp , pp, $\bar{p}$ p, ..) contained the same elementary constituents, the produced final states looked very different. The difference was even more striking when the comparison was made between processes involving just hadrons and those involving also leptons.

The observation that a pair of colliding hadrons (say pp) can "effectively" produce the same final states as a pair of annihilating leptons (say e^+e^-) thus came as a totally unexpected surprise. The key to explaining this apparent "miracle" was the concept of "effective energy". I will quote here a first remark by V.N. Gribov: *"When I read the paper "Evidence of the same multiparticle production mechanism in pp collisions as in e^+e^- annihilation" by A. Zichichi and collaborators, working with the pp ISR collider at CERN, I realized that something very interesting had been found. In fact the introduction of the "effective energy" in the analysis of pp collisions at CERN's ISR gave a totally unexpected result."*

The paper by A. Zichichi and collaborators [1], published in 1980, used data obtained from pp interactions at the Intersecting Storage Rings (ISR), where pairs of protons with total incident energy E = 62 GeV were colliding to produce low p_T multiparticle hadronic systems. Using a new kinematical variable — the effective energy — Zichichi's group was able to show that the ISR data at the high "nominal energy" of 62 GeV were in very good agreement with the e^+e^- data at much lower energies. In order to show this, individual pp events were classified according to their own *"effective energy"* which, in the comparison, had to be identified with the nominal energy of the e^+e^- data.

Here is a sentence from this paper [1] *"The agreement between the momentum distributions obtained in e^+e^- annihilation and in pp collisions suggests that the mechanism for transforming energy into particles in these two processes, so far considered very different, must be the same".*

Paper [1] was followed by a series of other results, all confirming the validity of this discovery. It was indeed a great surprise to see that introducing the effective energy, i.e. the energy effectively available in each reaction for the production of secondary particles, allowed to expose universal features even when the incoming particles were not hadronic, as in the two abovementioned examples, vp and e^+e^-. Let me quote here a sentence from another paper [2] of Zichichi's collaboration: "... *it does not matter whether the hadron interacts strongly, weakly, or electromagnetically: its «leading» effect is always present.*".

This work was very timely, as several difficulties were piling up when confronting QCD with experimental data (e.g. the η-η' problem, see below). These difficulties contrasted with its initial successes, when a new intrinsic degree of freedom (colour) was introduced to explain how totally symmetric SU(3)-flavour states, the well known baryons

$$(N^*)^{++}_{3/2,\,3/2} \quad \text{and} \quad \Omega^- ,$$

could be antisymmetric in the constituent fermions. The elementary constituents of all mesons and baryons thus needed to be interacting quarks and gluons.

Another remark by Gribov is appropriate here: "*In the physics community there was a sort of gentlemen's agreement: please do not speak about results in contrast with the so much searched for gauge interaction to describe hadronic phenomena. These "hidden" results were the hadronic systems produced in the interactions between pairs of hadrons; they were all different. Each pair of interacting particles, when producing systems consisting of many hadronic particles, had its own final state. No-one knew how to settle this flagrant contradiction. I wish I had the idea of the "effective energy".*"

The introduction of the "effective energy" concept allowed Nino to show that pairs of interacting particles, whether the interaction is strong, electromagnetic or weak, produce multiparticle hadronic states with the same basic properties. This provides an essential experimental support to the fundamental QCD process that is believed to lie at the basis of hadron production: the hadronization of quarks and gluons.

This section of the volume starts with the reproduction of the paper [1] that introduced the concept of "effective energy", and goes on to those that presented a systematic study of the multiparticle systems produced in strong interactions [3]: πp, pp, p̄p, Kp. Other very interesting results follow, where the introduction of the "effective energy" is shown to be needed also in processes induced by electromagnetic interactions (e^+e^-, γp, ep, μp) and in weak (νp) interactions, including deep inelastic scattering [4]. In fact, the leading effect is there whenever there is a "flow" of quantum numbers from the initial to the final state [2], no matter whether the interaction is strong, electromagnetic or weak.

Here is another statement from the same paper [2]: "*This supports the idea that the «leading» phenomenon is generated by the quantum number «flow» from the initial to the final state*". A posteriori, Zichichi's idea can be easily understood: if all hadronic showers originate from the fragmentation of quarks and gluons there has to be some universality. However, before being able to unravel it, one has to get rid of whatever features of the fragmenting partons are reaction-dependent. The leading effect and the concept of effective energy are good ways to take out (one of) the most important initial-state dependence. In other words, after the leading particle is produced and its energy is adequately taken out, the remaining system is much more universal than it was before, since any specific initial quantum number is eaten up by the final leading hadron (together with some energy, of course). Obviously, this is easy to say now, after the facts: one has to pay tribute to Nino's ingenuity for having had the right intuition so early on.

Let me elaborate further on my present understanding of this issue in the context of the so-called topological expansion (TE) of QCD, a systematic way of bookeeping QCD diagrams according to their topology. The leading diagrams have the simplest (planar) topology while increasingly non-planar topologies are considered as corrections. The expansion parameter turns out to be $1/N^2$, where N is the number of either the colours or the flavours, hence $O(1/10)$ which is hoped to be sufficiently small for the expansion to be meaningful.

In the TE, later developed into what has become known as the dual-parton model of high energy hadronic reactions, the final-state hadrons come grouped in jets. Each jet is the result of the pulling apart of a colour-singlet quark-antiquark pair, of the corresponding formation of a QCD string (a colour flux tube), and of its eventual break-up into the final-state mesons.

Does the Topological Expansion (TE) picture explain Nino's universality? My answer is that it does to some extent but with deviations from universality to be expected. The TE neatly identifies the relevant energy with Nino's effective energy. This should remove the main obstacle to universality.

Coming back to Nino's work, among the most interesting results I will mention here the proof that the two hemispheres in pp interactions at the ISR are independent [5] and the fact that it took five years, and the advent of PETRA, to prove that there is a leading effect also in e^+e^- jets, e.g. when charmed quarks represent the "initial" partonic state [6].

The review paper "The end of a myth: high$-p_T$ physics" [7] focuses the attention on the fact that the hadronization process is the key element in understanding the multihadronic final states. It is relatively unimportant whether these states are produced at high or low transverse momentum, provided they have the same "effective energy". Quoting the introduction from this paper: *"So far, the main picture of hadronic physics has been based on a distinction between high$-p_T$ and low$-p_T$ phenomena. In the framework of the parton model, high$-p_T$ processes were the only candidates to establish a link* *between*

- *purely hadronic processes*
- *(e^+e^-) annihilations*
- *(DIS) processes.*

The advent of QCD has emphasized in a dramatic way the privileged role of high$-p_T$ physics due to the fact that, thanks to asymptotic freedom, QCD calculations via perturbative methods can be attempted at high$-p_T$ and results successfully compared with experimental data. The conclusion was: we can forget about everything else and limit ourselves to high$-p_T$ physics.

Being theoretically off limits, low$-p_T$ phenomena, which represent the overwhelming majority of hadronic processes (more than 90% of physics being there), have been up to now neglected. By subtracting the leading proton effects in order to derive the effective energy available for particle production and by using the correct variables, the BCF collaboration has performed a systematic study of the final states produced in low$-p_T$ (pp) interactions at the ISR and has compared the results with those obtained in the processes listed below:

<u>Process</u>		<u>Data Sources</u>
(e^+e^-)		SLAC, DORIS, PETRA
(DIS)		SPS/EMC
(pp) ⎤		⎡ ISR (AFS)
⎥ *Transverse physics* ⎢		
($\bar{p}p$) ⎦		⎣ SPS Collider (UA1)
(e^+e^-)		PETRA/TASSO
		(leading effect in D* production).

The results of this study [2-18]* show that, once a common basis for

---

*   *These references can be found in the Ref. 7 reproduced in the present volume.*

comparison is found by the use of the correct variables, remarkable analogies are observed in processes so far considered basically different like

- *low–p_T (pp) interactions*
- *(e^+e^-) annihilations*
- *(DIS) processes*
- *high–p_T (pp) and ($\bar{p}$p) interactions*

This is how universality features emerge."

A very short synthesis of the BCF results has been given during the 1996 Erice School by A. Zichichi [8] with the purpose of calling the attention of QCD theorists on the need to describe these non perturbative QCD phenomena. At that time there was one missing piece of evidence, namely the leading effect in gluon induced jets.

The papers reproduced in this section of the present volume are of considerable interest in that they represent the "hidden" side of QCD physics, which no-one knew (and even now knows) how to reduce to a common basis. The concept of "effective energy" was needed, but this quantity was not simple to disentangle in the analysis of multihadronic final states. The first paper of this series [1] was in fact the result of a detailed study of many thousands of pp collisions, each one measured in detail. The nominal energy $(\sqrt{s})_{pp}$ was 62 GeV, but the effective energies covered a continuous spectrum. As illustrated in paper [9], other nominal ISR energies were also studied, each giving its own continuous spectrum of "effective energy". I would like to quote a few sentences from the review paper [10]: *"One can produce multiparticle hadronic states in basically three ways: i) hadron-hadron interactions; ii) e^+e^- annihilation; iii) lepton-hadron deep inelastic scattering (DIS). As mentioned above, the only way, so far, to correlate the hadronic systems* *produced in these different ways was to compare the properties, measured in e^+e^- and (DIS), with high$-p_T$ hadron-hadron data. The low$-p_T$ data were considered as examples of typically hadronic phenomena, not suitable for comparison with e^+e^- and DIS physics. Our new way of analysing pp interactions[1-12**] brings into the exciting field of physics the large amount of, so far abandoned, low$-p_T$ pp data.*

*The study of the properties of multihadronic systems produced in low$-p_T$ proton-proton interactions, once the energy for particle production in a pp interaction has been correctly calculated, shows that a striking series of analogies can be established between pp and e^+e^- processes[1-12**]".*

The last paper of this series [11] is the first evidence to fill the previously mentioned gap in the leading-effect picture, i.e. the existence of the leading effect in gluon induced jets as an energetic η' particle. This idea appears now to be substantiated by other observations in the decay of heavy-quark systems. It also fits well with the commonly accepted QCD explanation of the η–η' puzzle, i.e. that the difference in their masses is due to a large gluonic component in the latter's wavefunction.

Probably, the best way to conclude this report is by quoting once more Professor Gribov [12]: "*Think of it. Even after so many years it appears to me a great achievement in physics*". Gribov's dream was to formulate an asymptotically free theory in such a way that "*it includes perturbative and non-perturbative phenomena simultaneously*". When QCD will be formulated as Gribov was hoping, the leading effect and the universal features of multihadronic systems will probably become an integral part of QCD "predictions".

** *These references can be found in the review paper [10] reproduced in the present volume.*

REFERENCES

[1] *Evidence of the Same Multiparticle Production Mechanism in p-p Collisions as in e^+e^- Annihilation*
M. Basile, G. Cara Romeo, L. Cifarelli, A. Contin, G. D'Alì, P. Di Cesare, B. Esposito, P. Giusti, T. Massam, F. Palmonari, G. Sartorelli, G. Valenti and A. Zichichi
Physics Letters <u>92B</u>, 367 (1980).

[2] *The "Leading"-Baryon Effect in Strong, Weak, and Electromagnetic interactions*
M. Basile, G. Cara Romeo, L. Cifarelli, A. Contin, G. D'Alì, P. Di Cesare, B. Esposito, P. Giusti, T. Massam, R. Nania, F. Palmonari, V. Rossi, G. Sartorelli, M. Spinetti, G. Susinno, G. Valenti, L. Votano and A. Zichichi
Lettere al Nuovo Cimento <u>32</u>, 321 (1981).

[3] *The "Leading"-Particle Effect in Hadron Physics*
M. Basile, G. Cara Romeo, L. Cifarelli, A. Contin, G. D'Alì, P. Di Cesare, B. Esposito, P. Giusti, T. Massam, R. Nania, F. Palmonari, V. Rossi, G. Sartorelli, M. Spinetti, G. Susinno, G. Valenti, L. Votano and A. Zichichi
Nuovo Cimento <u>66A</u>, 129 (1981).

[4] *Universality Features in (pp), (e^+e^-) and Deep-Inelastic-Scattering Processes*
M. Basile, G. Bonvicini, G. Cara Romeo, L. Cifarelli, A. Contin, M. Curatolo, G. D'Alì, C. Del Papa, B. Esposito, P. Giusti, T. Massam, R. Nania, F. Palmonari, G. Sartorelli, G. Susinno, L. Votano and A. Zichichi
Nuovo Cimento <u>79A</u>, 1 (1984).

[5] *Experimental Proof that the Leading Protons are not Correlated*
M. Basile, G. Bonvicini, G. Cara Romeo, L. Cifarelli, A. Contin, M. Curatolo, G. D'Alì, C. Del Papa, B. Esposito, P. Giusti, T. Massam, R. Nania, F. Palmonari, G. Sartorelli, G. Susinno, L. Votano and A. Zichichi
Nuovo Cimento <u>73A</u>, 329 (1983).

[6] *Universality Properties in non-Perturbative QCD - Leading D^* in (e^+e^-) at PETRA*
A. Zichichi
Proceedings of the XXIII Course of the "Ettore Majorana" International School of Subnuclear Physics, Erice, Italy, 4-14 August 1985: "Old and New Forces of Nature" (Plenum Press, New York-London, 1988), 117.

[7]   *The End of a Myth: High-P_T Physics*
      M. Basile, J. Berbiers, G. Cara Romeo, L. Cifarelli, A. Contin, G. D'Alì,
      C. Del Papa, P. Giusti, T. Massam, R. Nania, F. Palmonari, G. Sartorelli,
      M. Spinetti, G. Susinno, L. Votano and A. Zichichi
      Opening Lecture in Proceedings of the XXII Course of the "Ettore Majorana"
      International School of Subnuclear Physics, Erice, Italy, 5-15 August 1984:
      "Quarks, Leptons, and their Constituents" (Plenum Press, New York-London,
      1988), 1.

[8]   *Universality Features in Multihadronic Final States*
      Old but still interesting results on the properties of multihadronic final states
      produced in Strong, Electromagnetic and Weak Interactions: a brief recalling
      A. Zichichi
      Proceedings of the 34th Course of the "Ettore Majorana" International School of
      Subnuclear Physics, Erice, Italy, 3-12 July 1996: "Effective Theories and
      Fundamental Interactions" (World Scientific Publishing, 1997), 498.

[9]   *A Detailed Study of $\langle n_{ch} \rangle$ versus E^{had} and $m_{1,2}$ at Different $(\sqrt{s})_{pp}$ in (pp)*
      Interactions
      M. Basile, G. Bonvicini, G. Cara Romeo, L. Cifarelli, A. Contin, M. Curatolo,
      G. D'Alì, P. Di Cesare, B. Esposito, P. Giusti, T. Massam, R. Nania,
      F. Palmonari, A. Petrosino, V. Rossi, G. Sartorelli, M. Spinetti, G. Susinno,
      G. Valenti, L. Votano and A. Zichichi
      Nuovo Cimento 67A, 244 (1982).

[10] *What we can Learn from High-Energy, Soft (pp) Interactions*
      M. Basile, G. Bonvicini, G. Cara Romeo, L. Cifarelli, A. Contin, M. Curatolo,
      G. D'Alì, B. Esposito, P. Giusti, T. Massam, R. Nania, F. Palmonari,
      A. Petrosino, V. Rossi, G. Sartorelli, M. Spinetti, G. Susinno, G. Valenti,
      L. Votano and A. Zichichi
      Proceedings of the XIX Course of the "Ettore Majorana" International School of
      Subnuclear Physics, Erice, Italy, 31 July-11 August 1981: "The Unity of the
      Fundamental Interactions" (Plenum Press, New York-London, 1983), 695.

[11] *Evidence for η' Leading Production in Gluon-Induced Jets*
      Presented by A. Zichichi at the 35th Course of the "Ettore Majorana"
      International School of Subnuclear Physics, Erice, Italy, 26 August-4
      September 1997: "Highlights of Subnuclear Physics: 50 Years Later" (World
      Scientific Publishing, 1999), 474.

[12]   V. Gribov
in "Effective Theories and Fundamental Interactions", International School of Subnuclear Physics (Erice 1996), Vol. 34, World Scientific Publishing, p. 500.

THE ORIGIN OF THE EFFECTIVE ENERGY AND OF THE UNIVERSALITY FEATURES: REPRODUCTION OF THE PAPERS QUOTED BY G. VENEZIANO

CERN
SERVICE D'INFORMATION
SCIENTIFIQUE

Volume 92B, number 3,4   PHYSICS LETTERS   19 May 1980

EVIDENCE OF THE SAME MULTIPARTICLE PRODUCTION MECHANISM
IN p–p COLLISIONS AS IN e+e− ANNIHILATION

M. BASILE [b], G. CARA ROMEO [c], L. CIFARELLI [c], A. CONTIN [a], G. D'ALI [c],
P. DI CESARE [c], B. ESPOSITO [d], P. GIUSTI [c], T. MASSAM [c], F. PALMONARI [c],
G. SARTORELLI [c], G. VALENTI [a] and A. ZICHICHI [a]

[a] CERN, Geneva, Switzerland
[b] Istituto di Fisica dell'Università di Bologna, Bologna, Italy
[c] Istituto Nazionale di Fisica Nucleare, Bologna, Italy
[d] Istituto Nazionale di Fisica Nucleare, Laboratori Nazionali di Frascati, Frascati, Italy

Received 5 March 1980

The split-field magnet spectrometer at the CERN intersecting storage rings was used to measure, in p–p collisions at $\sqrt{s}$ = 62 GeV, the inclusive momentum distribution of the charged particles produced in the same hemisphere as the leading proton ($x > 0.4$). A new scaling variable was introduced in order to take into account baryon-number conservation effects in p–p interactions. It is shown that distributions in this variable are in good agreement with the momentum distribution of the hadrons produced in e+e− annihilation. The results suggest that the multiparticle production mechanism in p–p collisions is the same as in e+e− provided that the effects of baryon-number conservation are removed.

1. Introduction. The parton picture of hadrons provides a unifying framework for describing and relating different processes such as lepton deep inelastic scattering, hadron–hadron high-p_T reactions, and e+e− annihilation.

The bulk of the hadron–hadron interactions, which are mostly at low p_T, are still not understood. Recently, many attempts have been made [1] to relate the hadron longitudinal momentum distributions in the fragmentation region of the proton to the quark distributions inside the proton. These attempts are model dependent, relying on the assumption of a particular recombination function of the constituents to form a pion. Furthermore, they are limited to the tail of the x distribution ($x > 0.5$).

We have used a completely new approach to study the inclusive momentum distributions of charged particles produced in p–p reactions and to compare them as closely as possible with the results from e+e− annihilation.

The system of hadrons produced in a p–p interaction must have a baryon number equal to two, but in e+e− annihilation the baryon number is zero. To in-

vestigate whether it is just baryon-number conservation that makes the difference between the multiparticle distributions in the two kinds of interaction, we have attempted to remove the effects of baryon-number conservation by selecting events with a leading proton and redefining the fractional variables of the particles so that they may "forget" the existence of that proton.

2. Data collection and analysis. The experiment was performed at the CERN intersecting storage rings (ISR) at $\sqrt{s}$ = 62 GeV, using the split-field magnet (SFM) and its multiwire proportional chamber (MWPC) detector. A detailed description of the facility can be found elsewhere [2].

The detector was operated in the simplest possible triggering mode, which required two or more charged tracks anywhere in the chambers. Under these conditions the trigger efficiency is effectively 100%, and only corrections for individual track efficiencies have to be applied. These events were reconstructed, and only the tracks whose momentum was determined with an estimated precision $\delta p/p < 0.3$ were retained.

Events with a probable proton were then selected,

Volume 92B, number 3,4   PHYSICS LETTERS   19 May 1980

using the criterion that the fastest particle in either hemisphere centred on the direction of the beams should be positive and should have a value of x = $2|p_L|/\sqrt{s}$ between 0.4 and 0.8. Since no direct particle identification was available, the choice of the lower value of x was suggested by the over-all inclusive particle distributions. At $x = 0.4$ the π^+ production rate is comparable with that of the protons. As x increases, the ratio p/π^+ increases, and so the assumption that the leading positive particle is a proton becomes more accurate; a $\delta p/p \lesssim 8\%$ for this leading particle was also required. After this selection, 4149 of the available sample of 38883 "minimum bias" events remained.

All other particles in the same hemisphere were assumed to be the hadronic system associated with that proton. The total energy of this "associated hadronic system" was computed from the energy of the beam, $\sqrt{s}/2$, and the fractional longitudinal momentum of the proton, $x_{\text{proton}} = 2p_L/\sqrt{s}$:

$$E_{\text{had}} = \tfrac{1}{2}\sqrt{s}(1 - |x_{\text{proton}}|) .$$

The inclusive momentum distribution of the particles of the "associated hadronic system" was expressed in terms of a new fractional variable x_R^*, defined as

$$x_R^* = p/E_{\text{had}};$$

that is, dividing the momentum p of the particle by the energy of the hadronic system rather than by the energy of the beams. For transforming their momenta from the laboratory system to the p–p centre-of-mass system where x and x_R^* were evaluated, the particles were assumed to be pions. A correction for the acceptance of the detector for single tracks, including the effects of momentum resolution cuts, was applied.

3. *Results.* Fig. 1 shows the results for various bands of the energy E_{had} of the hadronic system. These inclusive distributions are normalized by dividing the number of particles per bin of x_R^* by the total number of events. Only statistical errors are shown. Systematic errors in the acceptance calculations are up to 20% of the reported values.

In fig. 2 we compare our data for the energy bands $E_{\text{had}} = 5{-}8$ GeV, $8{-}11$ GeV, and $14{-}16$ GeV, with the e⁺e⁻ data from Tasso at PETRA [3] at corresponding beam energies $E_{\text{beam}} = 6.5$ GeV, $8.5{-}11$ GeV, and

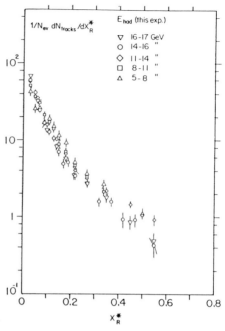

Fig. 1. Normalized single-particle inclusive distribution $(1/N_{\text{ev}}) (\text{d}N/\text{d}x_R^*)$ versus x_R^* for various bands of E_{had}, obtained from p–p collisions.

13.7–15.8 GeV. The quantity $s(\text{d}\sigma/\text{d}x_R)$ (with $x_R = 2p/\sqrt{s}$) given by the Tasso experiment has been multiplied by a factor $1/(2sR\sigma_{\mu\mu})$, where $\sigma_{\mu\mu}$ is the point-like μ cross section, and the values of $R = \sigma_{\text{had}}/\sigma_{\mu\mu}$ are taken from the same authors [4]. In this way we obtain the Tasso distributions normalized relative to only one half hemisphere, so that we can make an absolute comparison with our distributions as well as compare the shapes.

Figs. 2a,b and c show that the distribution as well as fractional momentum x_R^* of the hadrons produced in p–p collisions in our experiment is very similar, both in shape and absolute value (mean charged multiplicity), to the distribution of the fractional momentum of the hadrons produced in e⁺e⁻ annihilation.

4. *Conclusions.* The agreement between the momentum distributions obtained in e⁺e⁻ annihilation and p–p collisions suggests that the mechanism for transforming energy into particles in these two processes, so far considered very different, must be the

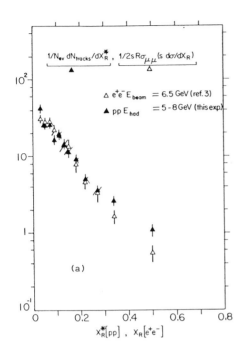

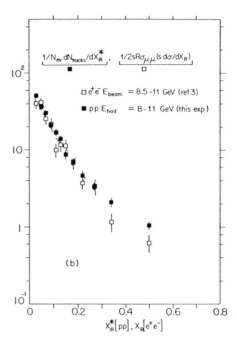

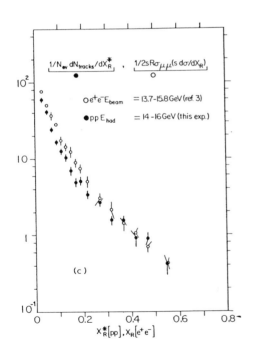

Fig. 2. The p–p single-particle inclusive distributions in terms of the variable x_R^*, are compared to the single-particle inclusive distributions obtained in e^+e^- annihilation; (a) E_{had} = 5–8 GeV, (b) E_{had} = 8–11 GeV, (c) E_{had} = 14–16 GeV.

same. In fact the main point of our work is to disentangle the kinematic effects of baryon-number conservation in p–p collisions. The results of our analysis could then provide a key to deeper understanding of the hadron–hadron reactions.

More studies, both experimental and theoretical, along these lines are certainly needed, and the results from the high-energy antiproton–proton colliders that are under construction will be particularly relevant.

The assistance of the SFMD group in the running of the detector was greatly appreciated. Finally we thank our technicians Messrs. J. Berbiers, F. Beauvais and P. Guerin, and Mmes. Y. Cholley and R. Blanc, for their continuous work.

Volume 92B, number 3,4　　　　　　　　PHYSICS LETTERS　　　　　　　　19 May 1980

References

[1] See, for example, K.P. Das and R.C. Hwa, Phys. Lett. 68B (1977) 459;
    M.J. Teper, RL-78-022/A (1978);
    D.W. Duke and F.E. Taylor, Phys. Rev. D17 (1978) 1788;
    F.C. Erné and J.C. Sens, Parton interpretation of low p_T inelastic p–p collisions at $s = 2000$ GeV^2, CERN preprint (1978), and references therein;
    L. Van Hove, Preprint TH. 2628-CERN (1979).

[2] R. Bouclier et al., Nucl. Instrum. Methods 125 (1975) 19.

[3] Tasso Collab., R. Brandelik et al., Phys. Lett. 89B (1980) 418; 83B (1979) 261.

[4] Tasso Collab., R. Brandelik et al., DESY 79/61 (1979).

LETTERE AL NUOVO CIMENTO   VOL. 32, N. 11   14 Novembre 1981

The « Leading »-Baryon Effect
in Strong, Weak, and Electromagnetic Interactions.

M. Basile, G. Cara Romeo, L. Cifarelli, A. Contin, G. D'Alí, P. Di Cesare,
B. Esposito, P. Giusti, T. Massam, R. Nania, F. Palmonari, V. Rossi, G. Sartorelli,
M. Spinetti, G. Susinno, G. Valenti, L. Votano and A. Zichichi

CERN - Geneva, Switzerland
Istituto di Fisica dell'Università - Bologna, Italia
Istituto Nazionale di Fisica Nucleare- Laboratori Nazionali di Frascati, Italia
Istituto Nazionale di Fisica Nucleare - Sezione di Bologna, Italia
Istituto di Fisica dell'Università - Perugia, Italia
Istituto di Fisica dell'Università - Roma, Italia

(ricevuto il 4 Settembre 1981)

Summary. – The « leading »-baryon effect is reported. A comparison is made between baryon-baryon and lepton-baryon interactions. The « leading » effect is present in both cases. This shows its relevance for the understanding of hadronic phenomena in strong, weak, and electromagnetic processes.

1. *Introduction.* – The purpose of the present paper is to report on a study of hadronic interactions induced on a baryon, either by another baryon or by a lepton. The main goal was to establish the existence of a « leading » effect when a hadron interacts strongly weakly, or electromagnetically.

In sect. 2 we define the « leading » quantity L; in sect. 3 the « leading » effect is studied in proton-proton collisions. Section 4 is devoted to the analysis of the lepton-baryon case. The conclusions are presented in sect. 5.

2. *Definition of the «leading» variable L.* – In a given hadronic reaction, the «leading» particle is the one carrying away a considerable fraction of the total available energy [1]. It is currently studied in terms of its fractional momentum variable:

$$(1) \qquad\qquad x = p_L/p_{L,max}, \qquad\qquad -1 \leqslant x \leqslant 1,$$

---

[1] M. Basile, G. Cara Romeo, L. Cifarelli, A. Contin, G. D'Alí, P. Di Cesare, B. Esposito, P. Giusti, T. Massam, R. Nania, F. Palmonari, V. Rossi, G. Sartorelli, M. Spinetti, G. Susinno, G. Valenti, L. Votano and A. Zichichi: *The « leading »-particle effect in hadron physics*, preprint CERN-EP/81-86, August 1981, *Nuovo Cimento*, in press.

322   M. BASILE, G. CARA ROMEO, L. CIFARELLI, A. CONTIN, G. D'ALÍ, ETC.

where p_L is the longitudinal-momentum component along the beam direction in the centre-of-mass system.

In order to quantify the « leading » effect, we have introduced the following variable:

$$(2) \qquad L(x_0, x_1, x_2) = \frac{\int_{x_1}^{x_2} F(x)\, dx}{\int_{x_0}^{x_1} F(x)\, dx} ,$$

where $F(x)$ is the inclusive single-particle cross-section. This is defined as

$$(3) \qquad F(x) = \frac{1}{\pi} \int \frac{2E}{\sqrt{s}} \frac{d^2\sigma}{dx\, dp_T^2}\, dp_T^2 ,$$

where E is the energy and p_T the transverse momentum of the particle, and $\sqrt{s}$ is the total centre-of-mass energy.

The integration limits used in formula (2) are $x_0 = 0.2$, $x_1 = 0.4$ and $x_2 = 0.8$. These are chosen in order to reduce diffractive ($x > 0.8$) and central ($x < 0.2$) production effects.

3. *The « leading »-baryon effect in proton-proton interactions.* – We have studied the quantity $L(x_0, x_1, x_2)$ for different hadrons produced in (pp) collisions at the CERN ISR [2-5]

(4a)   pp → p + anything ,

(4b)   pp → n + anything ,

(4c)   pp → Λ^0 + anything ,

(4d)   pp → Σ^+ + anything ,

(4e)   pp → Σ^- + anything ,

(4f)   pp → $\bar{p}$ + anything ,

(4g)   pp → $\overline{\Lambda^0}$ + anything .

The data refer to energies from 25 to 62 GeV. By inspecting the energy-dependence of L in the case of the neutron [4] (no data are available for other hadrons in a wide energy range), we conclude that the quantity L scales at ISR energies. It will be assumed to scale for all types of hadrons.

[1] P. CAPILUPPI, G. GIACOMELLI, A. M. ROSSI, G. VANNINI and A. BUSSIÈRE: *Nucl. Phys. B*, **70**, 1 (1974).
[2] M. G. ALBROW, A. BAGCHUS, D. P. BARBER, A. BOGAERTS, B. BOŠNJAKOVIĆ, J. R. BROOKS, A. B. CLEGG, F. C. ERNÉ, C. N. P. GEE, D. H. LOCKE, F. K. LOEBINGER, P. G. MURPHY, A. RUDGE and J. C. SENS: *Nucl. Phys. B*, **73**, 40 (1974).
[3] J. ENGLER, B. GIBBARD, W. ISENBECK, F. MÖNNIG, J. MORITZ, K. PACK, K. H. SCHMIDT, D. WEGENER, W. BARTEL, W. FLAUGER and H. SCHOPPER: *Nucl. Phys. B*, **84**, 70 (1975).
[4] S. EZHAN, W. LOCKMAN, T. MEYER, J. RANDER, P. SCHLEIN, R. WEBB and J. ZSEMBERY: *Phys. Lett. B*, **85**, 447 (1979).

In fig. 1 we have plotted the values of L derived from the experimental x distributions of the different hadrons in the reactions (4a) to (4g). The final states are ordered according to the number of quarks propagating from the initial state (the proton) to the final state.

The results of fig. 1 show that for different hadrons the value of L is the same, provided the number of propagating quarks is the same.

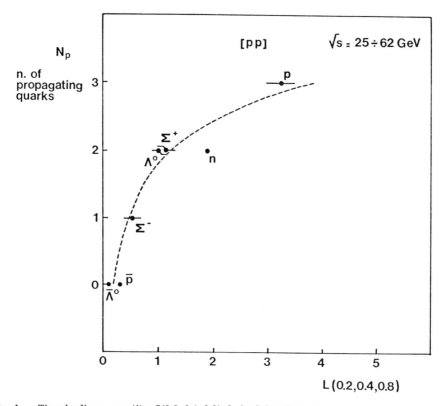

Fig. 1. – The «leading» quantity $L(0.2, 0.4, 0.8)$ derived for different types of baryons produced in (pp) collisions at the CERN ISR. The centre-of-mass energy ranges from 25 to 62 GeV. The hadrons are ordered according to the number of propagating quarks. The dashed curve superimposed is obtained by using a parametrization of the single-particle inclusive cross-section, $F(x) = (1 - x)^{\alpha}$, as described in sect. 3.

In order to compare the «leading» variable L with previous parametrizations ([6]), we have used in formula (2) the fit

$$(5) \qquad\qquad F(x) = (1 - x)^{\alpha},$$

where $\alpha = 2n - 1$ (n is the number of quarks which need to be changed in order to

([6]) J. S. GUNION: *Phys. Lett. B*, **88**, 150 (1979).

324   M. BASILE, G. CARA ROMEO, L. CIFARELLI, A. CONTIN, G. D'ALÍ, ETC.

obtain the wanted final hadron state). The dashed line in fig. 1 is the resulting variation of L vs. the number of propagating quarks, once the above parametrization (5) is used for $F(x)$.

The results show that the value of the « leading » quantity L is ~ 3 when the number of propagating quarks is 3. It is ~ 1.5 if the number of propagating quarks is two. It is ~ 0.5 when only one quark propagates, and it is < 0.5 when there is no quark propagation in the production of the final hadron state. Table I summarizes the results and describes the states in terms of their quark content.

TABLE I. – ISR (pp) *processes.*

Initial state: proton (udu)			
Final-state particle	Quark composition	Propagating quarks	$L(0.2, 0.4, 0.8)$
p	(udu)	(udu)	3.2 ± 0.2
n	(udd)	(ud)	1.92 ± 0.05
Λ^0	(uds)	(ud)	1.02 ± 0.10
Σ^+	(usu)	(uu)	1.15 ± 0.17
Σ^-	(dds)	(d)	0.53 ± 0.15
$\bar{p}$	$(\overline{u}\overline{d}\overline{u})$	nothing	0.30 ± 0.05
$\overline{\Lambda^0}$	$(\overline{u}\overline{d}\overline{s})$	nothing	0.10 ± 0.02

The « leading »-baryon effect is clearly present and it depends on the number of propagating quarks.

Recent studies on heavy-flavoured baryon produced in (pp) interactions at the CERN ISR provide evidence for the « leading » effect also in Λ_c^+ and Λ_b^0 production [7-11].

The next step is to investigate whether there is any « leading »-hadron effect when the hadron interacts either weakly or electromagnetically.

4. *The «leading»-baryon effect in lepton-baryon interactions (weak and electromagnetic). –* For this purpose we have studied the data on (vp) and (v̄p) interactions at Fermilab

[7] M. BASILE, G. CARA ROMEO, L. CIFARELLI, A. CONTIN, G. D'ALÍ, P. DI CESARE, B. ESPOSITO, P. GIUSTI, T. MASSAM, F. PALMONARI, G. SARTORELLI, G. VALENTI and A. ZICHICHI: *Nuovo Cimento A*, **63**, 230 (1981), and references therein.
[8] M. BASILE, G. CARA ROMEO, L. CIFARELLI, A. CONTIN, G. D'ALÍ, P. DI CESARE, B. ESPOSITO, P. GIUSTI, T. MASSAM, F. PALMONARI, G. SARTORELLI, G. VALENTI and A. ZICHICHI: *Lett. Nuovo Cimento*, **30**, 487 (1981).
[9] M. BASILE, G. BONVICINI, G. CARA ROMEO, L. CIFARELLI, A. CONTIN, G. D'ALÍ, P. DI CESARE, B. ESPOSITO, P. GIUSTI, T. MASSAM, R. NANIA, F. PALMONARI, G. SARTORELLI, G. VALENTI and A. ZICHICHI: *Lett. Nuovo Cimento*, **31**, 97 (1981).
[10] M. BASILE, G. BONVICINI, G. CARA ROMEO, L. CIFARELLI, A. CONTIN, G. D'ALÍ, P. DI CESARE, B. ESPOSITO, P. GIUSTI, T. MASSAM, R. NANIA, F. PALMONARI, G. SARTORELLI, G. VALENTI and A. ZICHICHI: *A comparison between « beauty » and « charm » production in* (pp) *interactions*, preprint CERN-EP/81-75, July 1981, *Nuovo Cimento*, in press.
[11] M. BASILE, M. BONVICINI, G. CARA ROMEO, L. CIFARELLI, A. CONTIN, G. D'ALÍ, P. DI CESARE, B. ESPOSITO, P. GIUSTI, T. MASSAM, R. NANIA, F. PALMONARI, G. SARTORELLI, G. VALENTI and A. ZICHICHI: *The « leading »-baryon effect in* Λ_b^0 *production in proton-interactions at* $\sqrt{s} = 62$ GeV, preprint CERN-EP/81-72, July 1981, *Nuovo Cimento*, in press.

energies ([12-14]). The proton x-distribution was measured in (νp) and (ν̄p) charged-current events, at the Fermilab 15 ft hydrogen bubble chamber ([12,13]). The most significant data refer to antineutrino interactions

(6) $\bar{\nu} + p \rightarrow \mu^+ + p + \text{anything}$,

at the energy $\langle W^2 \rangle \simeq 25 \,(\text{GeV})^2$ (where W is the invariant mass of the final-state hadronic system recoiling againts the muon). This distribution shows a very flat behaviour in the proton-recoil hemisphere ($x < 0$), thus establishing the existence of the « leading »-proton effect in weak processes. The value of $L(x_0, x_1, x_2)$ for reaction (6) is $L_p = 3.2 \pm 0.5$.

Further evidence for the « leading »-baryon effect in this kind of process can be derived from the inclusive Λ^0 production. The best data in this case are again those of the charged-current reaction ([14])

(7) $\bar{\nu} + \mathcal{N} \rightarrow \mu^+ + \Lambda^0 + \text{anything}$,

observed in the Fermilab 15 ft bubble chamber filled with a Ne-H$_2$ mixture, at $\langle W^2 \rangle \simeq$ $\simeq 20 \,(\text{GeV})^2$. The measured x-distribution of the Λ^0 exhibits a clean « leading » effect in the $x < 0$ region.

The value of $L(x_0, x_1, x_2)$ derived for reaction (7) is plotted in fig. 2. It agrees well with the corresponding L value obtained for reaction (4c) in pp interactions at higher energy, as shown in fig. 1. This study proves that the « leading »-baryon effect appears in weak interactions as firstly observed in purely hadronic interactions.

The same analysis can be repeated for the inclusive Λ^0 electroproduction:

(8) $e^- + p \rightarrow e^- + \Lambda^0 + \text{anything}$.

This was measured in a streamer chamber experiment at Cornell, in the energy range $5 < W^2 < 18 \,(\text{GeV})^2$ ([15]). The data again show that the Λ^0 is produced in a « leading » way even in electromagnetic interactions. The $L(x_0, x_1, x_2)$ value, relative to reaction (8), is only slightly higher than the analogous value found for reaction (7), as shown in fig. 2, and both these values compare well with the value from reaction (4c).

([12]) J. BELL, C. T. COFFIN, R. N. DIAMOND, H. T. FRENCH, W. C. LOUIS, B. P. ROE, R. T. ROSS, A. A. SEIDL, J. C. VANDER VELDE, E. WANG, J. P. BERGE, D. V. BOGERT, F. A. DiBIANCA, R. ENDORF, R. HABFT, C. KOCHOWSKI, J. A. MALKO, G. I. MOFFATT, F. A. NEZRICK, W. G. SCOTT, W. SMART, R. J. CENCE, F. A. HARRIS, M. JONES, M. W. PETERS, V. Z. PETERSON, V. J. STENGER, G. R. LYNCH, J. P. MARRINER and M. L. STEVENSON: Phys. Rev. D, 19, 1 (1979).
([13]) M. DERRICK, P. GREGORY, F. LoPINTO, B. MUSGRAVE, J. SCHLERETH, P. SCHREINER, R. SINGER, S. J. BARISH, R. BROCK, A. ENGLER, T. KIKUCHI, R. W. KRAEMER, F. MESSING, B. J. STACEY. M. TABAK, V. E. BARNES, T. S. CARMAN, D. D. CARMONY, E. FERNANDEZ, A. F. GARFINKEL and A. T, LAASANEN: Phys. Rev. D, 24, 1071 (1981).
([14]) V. AMMOSOV, A. AMRAKHOV, A. DENISOV, P. ERMOLOV, V. GAPIENKO, V. KLYUKHIN, V. KORESHOV, P. PITUKHIN, V. SIROTENKO, E. SLOBODYUK, V. ZAETZ, J. P. BERGE, D. BOGERT, R. HANFT, J. MALKO, G. HARIGEL, G. MOFFATT, F. NEZRICK, J. WOLFSON, V. EFREMENKO, A. FEDOTOV, P. GORICHEV, V. KAFTANOV, G. KLIGER, V. KOLGANOV, S. KRUTCHININ, M. KUBANTSEV, I. MAKHLYUEVA, V. SHE-KELJAN, V. SHEVCHENKO, J. BELL, C. T. COFFIN, W. LOUIS, B. P. ROE, R. T. ROSS, D. SINCLAIR and E. WANG: Nucl. Phys. B, 162, 205 (1980), and references therein.
([15]) I. COHEN, R. ERICKSON, F. MESSING, E. NORDBERG, R. SIEMANN, J. SMITH-KINTNER, P. STEIN, G. DREWS, W. GEBERT, P. JOOS, A. LADAGE, H. NAGEL, P. SÖDING and A. SADOFF: Phys. Rev. Lett., 40, 1614 (1978).

326   M. BASILE, G. CARA ROMEO, L. CIFARELLI, A. CONTIN, G. D'ALÍ, ETC.

It should be noticed, as discussed in ref. (^{1}), that for Λ^0 production the energy dependence of L shows the same features as those observed for proton production. In fact, the three values of L obtained for the Λ^0 at ISR, at Fermilab, and at Cornell energies, suggest the following trend: the higher is the available energy, the lower is the value of L.

All the above results thus point out that it does not matter whether the hadron interacts strongly, weakly, or electromagnetically: its « leading » effect is always present.

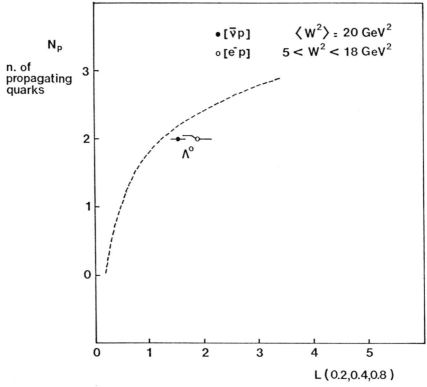

Fig. 2. – The « leading » quantity $L(0.2, 0.4, 0.8)$ of the Λ^0 produced in $(\bar{\nu}p)$ interactions at $\langle W^2 \rangle = 20$ (GeV)^2 and in (e$^-$p) interactions with $5 < W^2 < 18$ (GeV)^2. In this case the number of propagating quarks is two. The dashed curve of fig. 1 is also shown to guide the eye.

5. *Conclusions.* – In baryon-baryon interactions, the «leading»-baryon effect shows up very clearly in the x range $(0.2 \div 0.8)$. This « leading » effect is maximum when the final-state hadron is the same as the initial-state hadron. However, the « leading » effect is present even when the initial-state quantum numbers differ from those of the final state (for instance, when a proton becomes a Λ^0). As the difference between the initial- and the final-state quark composition increases, the « leading » effect decreases. This supports the idea that the « leading » phenomenon is generated by the quantum number « flow » from the initial to the final state.

The « leading » baryon effect appears both in baryon-baryon and in lepton-baryon interactions. This means that a definite similarity must exist between processes in which a hadron is present in the initial state, no matter if the interaction is strong, weak or electromagnetic.

CERN
SERVICE D'INFORMATION
SCIENTIFIQUE

M. BASILE, *et al.*
14 Novembre 1981
Lettere al Nuovo Cimento
Serie 2, Vol. 32, pag. 321-326

IL NUOVO CIMENTO   VOL. 66 A, N. 2   21 Novembre 1981

The « Leading »-Particle Effect in Hadron Physics.

M. BASILE, G. CARA ROMEO, L. CIFARELLI, A. CONTIN, G. D'ALÍ, P. DI CESARE,
B. ESPOSITO, P. GIUSTI, T. MASSAM, R. NANIA, F. PALMONARI, V. ROSSI,
G. SARTORELLI, M. SPINETTI, G. SUSINNO, G. VALENTI,
L. VOTANO and A. ZICHICHI

CERN - Geneva, Switzerland
Istituto di Fisica dell'Università - Bologna, Italia
Istituto Nazionale di Fisica Nucleare - Laboratori Nazionali di Frascati, Italia
Istituto Nazionale di Fisica Nucleare - Sezione di Bologna, Italia
Istituto di Fisica dell'Università - Perugia, Italia
Istituto di Fisica dell'Università - Roma, Italia

(ricevuto il 30 Luglio 1981)

Summary. — A world review is presented of the « leading »-particle effect in hadron physics: baryon-baryon, baryon-antibaryon, meson-baryon and photon-baryon processes are studied and compared with hadron production from an initial state which does not contain any hadron. In this latter case the « leading » effect is not present. It is shown that the « leading »-particle effect is fully present even in hadron interactions initiated by neutrinos. The importance of the « leading » effect in hadron physics is discussed.

1. – Introduction.

We have recently introduced a new method for studying hadronic reactions (¹). This method is based on the determination of the effective energy

---

(¹) A. ZICHICHI: *Evidence for a close link between* (pp) *and* (e⁺e⁻) *physics*, in *Proceedings of the « Ettore Majorana » International School of Subnuclear Physics, XVIII Course: The high-energy limit* (in press).

available for particle production. This allows, firstly, a comparison of the properties of multiparticle final states produced in purely hadronic interactions at different nominal c.m. energies. For example, in a proton-proton interaction at 62 GeV c.m. energy, the two protons can carry 40 GeV of the available energy. In this case the multiparticle system produced should be labelled by an effective energy of 22 GeV, and all quantities such as fractional-momentum distribution, average charged multiplicity, transverse-momentum distribution, etc., should be labelled with 22 GeV, not 62 GeV. Moreover, in another proton-proton interaction at 52 GeV c.m. energy, the two protons can carry 30 GeV of the available energy. In this case, again the multiparticle system produced should be labelled by the effective energy of 22 GeV. Thus two primary (pp) interactions, at $\sqrt{s} = 62$ and $\sqrt{s} = 52$ GeV, can produce identical final states, where the effective energy is 22 GeV. The key point is to subtract the energy taken away by the « leading » particles.

Another important consequence of our method is the extension of the comparison to hadronic final states which are not produced by the same hadrons. For example, (pp) interactions can be compared with (πp), (Kp) and (np) processes. Moreover, the comparison can be extended further: even to photon-induced and charged leptons plus neutrino-induced processes: (γp), (e$^\pm$p), (μ$^\pm$p), (νp), etc. Finally, the comparison can obviously be extended to final states produced in (e$^+$e$^-$) annihilation. The basic point is to take into account the effect of the « leading » hadron which, from the initial to the final state, keeps a highly privileged energy and momentum sharing. This « leading » four-momentum has to be correctly subtracted from the multiparticle final state produced.

This method has been applied extensively ([1-9]) in the study of multiparticle

([2])   M. BASILE, G. CARA ROMEO, L. CIFARELLI, A. CONTIN, G. D'ALÍ, P. DI CESARE, B. ESPOSITO, P. GIUSTI, T. MASSAM, F. PALMONARI, G. SARTORELLI, G. VALENTI and A. ZICHICHI: *Phys. Lett. B*, **92**, 367 (1980).
([3])   M. BASILE, G. CARA ROMEO, L. CIFARELLI, A. CONTIN, G. D'ALÍ, P. DI CESARE, B. ESPOSITO, P. GIUSTI, T. MASSAN, R. NANIA, F. PALMONARI, G. SARTORELLI, G. VALENTI and A. ZICHICHI: *Phys. Lett. B*, **95**, 311 (1980).
([4])   M. BASILE, G. CARA ROMEO, L. CIFARELLI, A. CONTIN, G. D'ALÍ, P. DI CESARE, B. ESPOSITO, P. GIUSTI, T. MASSAM, R. NANIA, F. PALMONARI, G. SARTORELLI, G. VALENTI and A. ZICHICHI: *Nuovo Cimento A*, **58**, 193 (1980).
([5])   M. BASILE, G. CARA ROMEO, L. CIFARELLI, A. CONTIN, G. D'ALÍ, P. DI CESARE, B. ESPOSITO, P. GIUSTI, T. MASSAM, R. NANIA, F. PALMONARI, G. SARTORELLI, G. VALENTI and A. ZICHICHI: *Lett. Nuovo Cimento*, **29**, 491 (1980).
([6])   M. BASILE, G. CARA ROMEO, L. CIFARELLI, A. CONTIN, G. D'ALÍ, P. DI CESARE. B. ESPOSITO, P. GIUSTI, T. MASSAM, R. NANIA, F. PALMONARI, G. SARTORELLI, M. SPINETTI, G. SUSINNO, G. VALENTI and A. ZICHICHI: *Phys. Lett. B*, **99**, 247 (1981),
([7])   M. BASILE, G. CARA ROMEO, L. CIFARELLI, A. CONTIN, G. D'ALÍ, P. DI CESARE, B. ESPOSITO, P. GIUSTI, T. MASSAM, R. NANIA, F. PALMONARI, G. SARTORELLI, M. SPINETTI, G. SUSINNO, G. VALENTI and A. ZICHICHI: *Lett. Nuovo Cimento*, **30**,

hadronic systems produced in low-p_T, high-energy, (pp) interactions. The « leading »-particle effect has also been applied to identify the presence of a high-energy proton, in order to search for new baryonic states ([10-14]) in (pp) interactions. In this study, without the « leading »-particle effect, it would be impossible to identify the high-energy proton.

The roots of this new approach to the study of hadronic interactions go back a long time, to a proposal by the CERN-Bologna Group to study « leptonic-like » (pp) collisions (CERN/PHI/COM-69/35, 8 July 1969).

The lesson that has been learnt is that, in a given hadronic reaction, it is in fact necessary to disentangle two phenomena,

    i) the « leading »-particle effect and

    ii) the multihadron production mechanism,

in order to understand the intrinsic features of both effects.

In this paper we present a review of the « leading »-particle phenomenon.

In sect. **2** we give a definition of the « leading » particle and of the variables used; in sect. **3** we show that in $e^+e^- \to$ hadrons, no « leading »-particle effect is present. Section **4** reviews the experimental evidence of the « leading »-particle effect in hadron-hadron interactions; the experimental evidence for the same phenomenon in lepton-hadron interactions is given in sect. **5**. Conclusions are finally given in sect. **6**.

389 (1981).

(8) M. BASILE, G. CARA ROMEO, L. CIFARELLI, A. CONTIN, G. D'ALÍ, P. DI CESARE, B. ESPOSITO, P. GIUSTI, T. MASSAM, R. NANIA, F. PALMONARI, G. SARTORELLI, G. VALENTI and A. ZICHICHI: *Lett. Nuovo Cimento*, **31**, 273 (1981).

(9) M. BASILE, G. CARA ROMEO, L. CIFARELLI, A. CONTIN, G. D'ALÍ, P. DI CESARE, B. ESPOSITO, P. GIUSTI, T. MASSAM, R. NANIA, F. PALMONARI, V. ROSSI, G. SARTORELLI, M. SPINETTI, G. SUSINNO, G. VALENTI, L. VOTANO and A. ZICHICHI: *Nuovo Cimento A*, **65**, 400 (1981).

(10) M. BASILE, G. CARA ROMEO, L. CIFARELLI, A. CONTIN, G. D'ALÍ, P. DI CESARE, B. ESPOSITO, P. GIUSTI, T. MASSAM, F. PALMONARI, G. SARTORELLI, G. VALENTI and A. ZICHICHI: *Lett. Nuovo Cimento*, **30**, 487 (1981).

(11) M. BASILE, G. CARA ROMEO, L. CIFARELLI, A. CONTIN, G. D'ALÍ, P. DI CESARE, B. ESPOSITO, P. GIUSTI, T. MASSAM, F. PALMONARI, G. SARTORELLI, G. VALENTI and A. ZICHICHI: *Nuovo Cimento A*, **63**, 230 (1981).

(12) M. BASILE, G. BONVICINI, G. CARA ROMEO, L. CIFARELLI, A. CONTIN, G. D'ALÍ, P. DI CESARE, B. ESPOSITO, P. GIUSTI, T. MASSAM, R. NANIA, F. PALMONARI, G. SARTORELLI, G. VALENTI and A. ZICHICHI: *Lett. Nuovo Cimento*, **31**, 97 (1981).

(13) M. BASILE, G. BONVICINI, G. CARA ROMEO, L. CIFARELLI, A. CONTIN, G. D'ALÍ, P. DI CESARE, B. ESPOSITO, P. GIUSTI, T. MASSAM, R. NANIA, F. PALMONARI, G. SARTORELLI, G. VALENTI and A. ZICHICHI: *Nuovo Cimento A*, **65**, 391 (1981).

(14) M. BASILE, G. BONVICINI, G. CARA ROMEO, L. CIFARELLI, A. CONTIN, G. D'ALÍ, P. DI CESARE, B. ESPOSITO, P. GIUSTI, T. MASSAM, R. NANIA, F. PALMONARI, G. SARTORELLI, C. VALENTI and A. ZICHICHI: *Nuovo Cimento A*, **65**, 408 (1981).

2. – Definition of the « leading » quantity « L ».

Consider the inclusive reaction

$$(1) \qquad \qquad a+b \rightarrow c+\text{anything} .$$

Particle c is a « leading » particle in reaction (1) if, on the average, it carries a sizable fraction of the total energy available. The most convenient way of studying the « leading »-particle effect in reaction (1) is, therefore, to use fractional energy or momentum variables, generally labelled by the symbol x.

Let E_c and p_c be the energy and momentum of particle c in the final state; E_{max} and p_{max} the kinematic bounds for E_c and p_c in reaction (1). The currently used fractional variables are x_F and x_R. The first variable has been introduced by FEYNMAN,

$$(2) \qquad \qquad x_F = p_L/p_{L,max} , \qquad \qquad -1 \leqslant x_F \leqslant 1,$$

where p_L is the momentum component of particle c along the projectile direction in the centre-of-mass system for reaction (1); x_F ranges from -1 to $+1$, the negative values corresponding to the so-called target x-region, the positive values to the projectile x-region.

The radial variable x_R is the fractional energy defined according to

$$(3) \qquad \qquad x_R = E_c/E_{max} , \qquad \qquad 0 \leqslant x_R \leqslant 1 .$$

For the sake of clarity, the type of x variable used will be explicitly indicated throughout the paper.

The inclusive single-particle cross-section integrated over p_T, used in this paper and referred to as $F(x)$, is defined according to

$$(4) \qquad \qquad F(x) = \frac{1}{\pi} \int \frac{2E_c}{\sqrt{s}} \frac{\mathrm{d}^2\sigma}{\mathrm{d}x\,\mathrm{d}p_T^2} \,\mathrm{d}p_T^2 ,$$

where $\sqrt{s}$ is the total centre-of-mass energy.

In order to quantify the « leading »-hadron effect, we have defined the following quantity:

$$(5) \qquad \qquad L(x_0, x_1, x_2) = \frac{\int_{x_1}^{x_2} F(x)\,\mathrm{d}x}{\int_{x_0}^{x_1} F(x)\,\mathrm{d}x} ,$$

where $F(x)$ is the experimental invariant cross-section defined in eq. (4). The ranges of x_F used are: $x_0 = 0.2$, $x_1 = 0.4$, $x_2 = 0.8$. They have been chosen in order to minimize the effect of the diffractive production ($x_F > 0.8$) and of the central production ($x_F < 0.2$).

THE «LEADING»-PARTICLE EFFECT IN HADRON PHYSICS   **133**

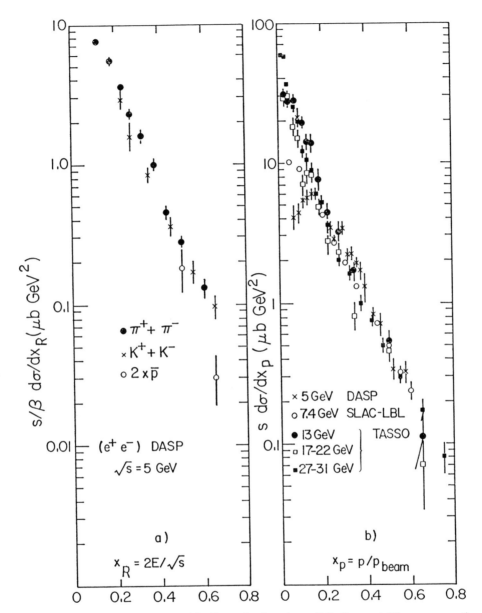

Fig. 1. – Inclusive production of hadrons in (e⁺e⁻) annihilation. *a*) The cross-section $(s/\beta)(d\sigma/dx_R)$ *vs.* x_R for $(\pi^+ + \pi^-)$, $(K^+ + K^-)$ and twice the $\bar{p}$ yield for $\sqrt{s} = 5$ GeV (ref. ([15])). *b*) The cross-section $s(d\sigma/dx_p)$ for inclusive charged-particle production measured at energies of 5 to 30 GeV (ref. ([16])).

([15]) DASP COLLABORATION (R. BRANDELIK *et al.*): *Phys. Lett. B*, **67**, 358 (1977).
([16]) G. WOLF: DESY 80/13 (1980). Quoted data are taken from i) DASP COLLABORATION (R. BRANDELIK *et al.*): *Nucl. Phys. B*, **148**, 189 (1979); ii) G. J. FELDMAN and M. L. PERL: *Phys. Rep.*, **33**, 285 (1977); iii) TASSO COLLABORATION (R. BRANDELIK *et al.*): *Phys. Lett. B*, **89**, 418 (1980).

134   M. BASILE, G. CARA ROMEO, L. CIFARELLI, A. CONTIN, G. D'ALÍ, ETC.

3. – Hadron production in (e⁺e⁻) annihilation: no « leading »-hadron effect.

The question we ask ourselves is: Given a certain amount of energy to be transformed into hadrons, does the produced multiparticle final-state system have a « leading » hadron? The answer comes from (e⁺e⁻) annihilation experiments: according to the experimental results obtained up to the highest energies studied so far, (e⁺e⁻) annihilation does not produce hadronic systems with a « leading » hadron. This is shown in the data reported below.

Figure 1a) shows the inclusive cross-section $(s/\beta)(\mathrm{d}\sigma/\mathrm{d}x_\mathrm{R})$ vs. $x_\mathrm{R} = 2E/\sqrt{s}$

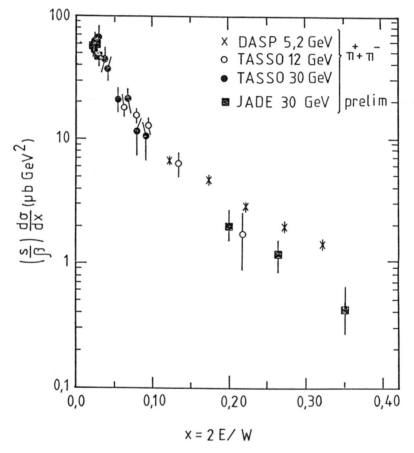

Fig. 1. – Inclusive production of hadrons in (e⁺e⁻) annihilation. c) The scaled cross-section $(s/\beta)\,\mathrm{d}\sigma/\mathrm{d}x$ for charged pions (ref. ([17,18])).

([17]) D. PANDOULAS: *Proceedings of the XX International Conference on High-Energy Physics, Madison, Wis., 1980* (New York, N. Y., 1981), p. 596.
([18]) TASSO COLLABORATION (R. BRANDELIK et al.): *Phys. Lett. B*, **94**, 444 (1980).

for $\pi^\pm$, $K^\pm$ and $\bar{p}$ measured at DORIS by the DASP Collaboration [15] at $\sqrt{s} = 5$ GeV. The experimental points for π, K and $\bar{p}$ have the same x_R distribution, peaked at $x_R = 0$ and falling off steeply at a large value of x_R. No hadron shows a «leading» effect.

Figure 1b) shows a compilation [16] of the inclusive cross-section $s(d\sigma/dx_p)$ vs. $x_p = |\mathbf{p}|/|\mathbf{p}_{\text{beam}}|$ for charged particles with $\sqrt{s}$ ranging from 5 to 30 GeV. The scaled cross-sections $(s/\beta)(d\sigma/dx)$ vs. $x = 2E/W$ are shown for charged pions [17,18] in fig. 1c), for charged and neutral kaons [17-19] in fig. 1d) and for

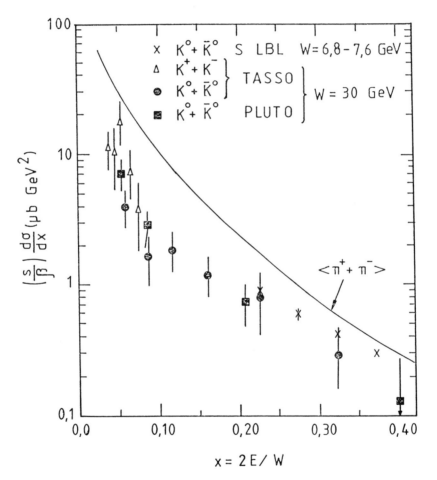

Fig. 1. – Inclusive production of hadrons in (e⁺e⁻) annihilation. d) The scaled cross-section $(s/\beta)\,d\sigma/dx$ for neutral and charged kaons (ref. [17-19]). The average pion cross-section is shown as the solid curve.

[19] V. Lüth, F. M. Pierre, G. S. Abrams, M. S. Alam, A. M. Boyarski, M. Breidenbach, W. Chinowsky, J. Dorfan, G. J. Feldman, G. Goldhaber, G. Hanson, J. A. Jaros, J. A. Kadyk, R. R. Larsen, A. Litke, D. Lüke, H. L. Lynch, R. J.

136   M. BASILE, G. CARA ROMEO, L. CIFARELLI, A. CONTIN, G. D'ALÍ, ETC.

protons and antiprotons ([18,20]) in fig. 1e). The distributions in fig. 1c), d) and e) are similar to the one in fig. 1a). They all clearly show a trend that is opposite to the « leading »-particle effect. This is quantitatively expressed by the values

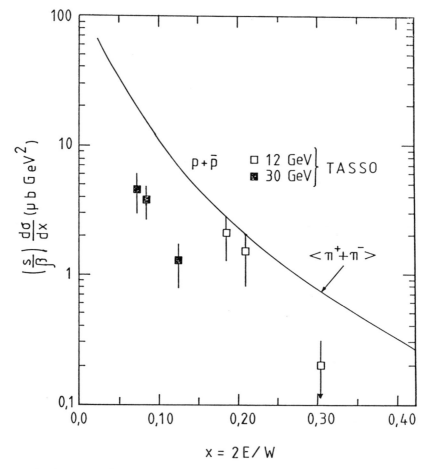

Fig. 1. – Inclusive production of hadrons in (e⁺e⁻) annihilation. e) The scaled cross-section $(s/\beta)\, d\sigma/dx$ for protons and antiprotons (ref. ([18,20])). The average charged-pion cross-section is shown as the solid curve.

of L which are always below 0.5. A summary of all L values is shown in table I. The value of « L » has been calculated on the distributions of fig. 1a and 1e).

MADARAS, H. K. NGUYEN, J. M. PATERSON, M. L. PERL, I. PERUZZI, CH. PEYROU, M. PICCOLO, T. P. PUN, P. RAPIDIS, B. RICHTER, B. SADOULET, R. H. SCHINDLER, R. H. SCHWITTERS, J. SIEGRIST, W. TANENBAUM, G. H. TRILLING, F. VANNUCCI, J. S. WHITAKER and J. E. WISS: *Phys. Lett. B*, **70**, 120 (1977).
([20]) TASSO COLLABORATION (R. BRANDELIK *et al.*): *Phys. Lett. B*, **89**, 418 (1980).

TABLE I. – *Value of the «leading» quantity «L» measured for various types of hadrons, i.e. π^+, π^-, K^+, K^-, p, $\bar{p}$, produced in e^+e^- interactions.*

Reaction	«L»
$e^+e^- \to \pi^+(\pi^-)$ +any	0.3 ± 0.1
$e^+e^- \to K^+(K^-)$+any	0.3 ± 0.1
$e^+e^- \to p(\bar{p})$ +any	0.3 ± 0.2

4. – The «leading»-hadron effect in purely hadronic interactions.

The following processes have been studied at ISR energies ([21-24]):

$$(A) \quad \begin{cases} [1] & pp \to p + \text{anything}, \\ [2] & pp \to n + \text{anything}, \\ [3] & pp \to \Lambda + \text{anything}, \\ [4] & pp \to \Sigma^+ + \text{anything}, \\ [5] & pp \to \Sigma^- + \text{anything}, \\ [6] & pp \to \bar{p} + \text{anything}, \\ [7] & pp \to \bar{\Lambda} + \text{anything}. \end{cases}$$

In order to compare those data with lower energy, we have studied the data taken on a fixed target (hydrogen) at incident-proton momenta of 100 GeV/c. Moreover, we have extended our study to other incident particles and even lower energies. This comparison is important for two reasons: i) it allows us

([21]) J. ENGLER, B. GIBBARD, W. ISENBECK, F. MÖNNING, J. MORITZ, K. PACK, K. H. SCHMIDT, D. WEGENER, W. BARTEL, W. FLAUGER and H. SCHOPPER: *Nucl. Phys. B,* **84**, 70 (1975).

([22]) P. CAPILUPPI, G. GIACOMELLI, A. M. ROSSI, G. VANNINI and A. BUSSIÈRE: *Nucl. Phys. B,* **70**, 1 (1974).

([23]) M. G. ALBROW, A. BAGCHUS, D. P. BARBER, A. BOGAERTS, B. BOSNJAKOVIC, J. R. BROOKS, A. B. CLEGG, F. C. ERNÈ, C. N. P. GEE, D. H. LOCKE, F. K. LOEBINGER, P. G. MURPHY, A. RUDGE and J. C. SENS: *Nucl. Phys. B,* **73**, 40 (1973).

([24]) S. ERHAN, W. LOCKMAN, T. MEYER, J. RANDER, P. SCHLEIN, R. WEBB and J. ZSEMBERY: *Phys. Lett. B,* **85**, 447 (1979).

138   M. BASILE, G. CARA ROMEO, L. CIFARELLI, A. CONTIN, G. D'ALÍ, ETC.

to measure the energy dependence of the quantity « L »; ii) it allows us to see whether « L » varies when different hadrons are used (π, K, p, $\bar{p}$, etc.).   The reactions investigated at 100 GeV/c are ([25-29]):

(B)

$$
\begin{cases}
[8] & \text{pp} \rightarrow \text{p+anything,} \\
[9] & \text{pp} \rightarrow \bar{\text{p}}\text{+anything,} \\
[10] & \text{pp} \rightarrow \pi^{+}\text{+anything,} \\
[11] & \text{pp} \rightarrow \pi^{-}\text{+anything,} \\
[12] & \text{pp} \rightarrow \text{K}^{+}\text{+anything,} \\
[13] & \text{pp} \rightarrow \text{K}^{-}\text{+anything,}
\end{cases}
$$

(C)

$$
\begin{cases}
[14] & \bar{\text{p}}\text{p} \rightarrow \bar{\text{p}}\text{+anything,} \\
[15] & \bar{\text{p}}\text{p} \rightarrow \pi^{+}\text{+anything,} \\
[16] & \bar{\text{p}}\text{p} \rightarrow \text{p+anything,}
\end{cases}
$$

(D)

$$
\begin{cases}
[17] & \pi^{-}\text{p} \rightarrow \pi^{-}\text{+anything,} \\
[18] & \pi^{-}\text{p} \rightarrow \pi^{+}\text{+anything,} \\
[19] & \pi^{-}\text{p} \rightarrow \text{p+anything.}
\end{cases}
$$

([25]) A. E. BRENNER: *Proceedings of the Seoul Symposium on Elementary Particles, Seoul, Korea, 1978* (Seoul, 1979), p. 709.

([26]) J. R. JOHNSON, R. KAMMERUD, T. OHSUGI, D. J. RITCHIE, R. SHAFER, D. THERIOT, J. K. WALKER and F. E. TAYLOR: *Phys. Rev. D*, **17**, 1292 (1978).

([27]) C. P. WARD, D. R. WARD, R. E. ANSORGE, J. R. CARTER, W. W. NEALE, J. G. RUSHBROOKE, B. Y. OH, M. PRATAP, G. A. SMITH, J. WHITMORE, R. RAJA and L. VOYVODIC: *Nucl. Phys. B*, **153**, 299 (1979).

([28]) J. WHITMORE, B. Y. OH, M. PRATAP, G. SIONAKIDES, G. A. SMITH, V. E. BARNES, D. D. CARMONY, R. S. CHRISTIAN, A. F. GARFINKEL, W. M. MORSE, L. K. RANGAN, L. VOYVODIC, R. WALKER, E. W. ANDERSON, H. B. CRAWLEY, A. FIRESTONE, W. J. KERNAN, D. L. PARKER, R. G. GLASSER, D. G. HILL, M. KAZUNO, G. McCLELLAN, H. L. PRICE, B. SECHI-ZORN, G. A. SNOW, F. SVRECK, A. R. ERWIN, E. H. HARVEY, R. J. LOVELESS and M. A. THOMPSON: *Phys. Rev. D*, **16**, 3137 (1977).

([29]) P. D. HIGGINS, N. N. BISWAS, J. M. BISHOP, R. L. BOLDUC, N. M. CASON, V. P. KENNEY, R. C. RUCHTI, W. D. SHEPHARD, W. D. WALKER, J. S. LOOS, L. R. FORTNEY, A. T. GOSHAW, W. J. ROBERTSON, E. W. ANDERSON, H. B. CRAWLEY, A. FIRESTONE, R. FLOYD, W. J. KERNAN, J. W. LAMSA, D. L. PARKER, G. A. SNOW, B. Y. OH, M. PRATAP, G. SIONAKIDES, G. A. SMITH, J. WHITMORE, V. SREEDHAR, G. LEVMAN, B. M. SCHWARZSCHILD, T. S. YOON, G. HARTNER, P. M. PATEL, L. VOYVODIC and R. J. WALKER: *Phys. Rev. D*, **19**, 731 (1979).

At 70 GeV/c of beam momentum the following reactions have been studied ([30,31]):

(E)
$$\begin{cases} [20] \quad \mathrm{K^+p \to c^+ + anything} \quad (\text{c}^+ \text{ is a positively} \\ \qquad\qquad\qquad\qquad\qquad\qquad\quad \text{charged particle}) \\ [21] \quad \mathrm{K^+p \to \pi^- + anything,} \\ [22] \quad \mathrm{K^+p \to p + anything.} \end{cases}$$

The following reactions:

(F)
$$\begin{cases} [23] \quad \mathrm{K^-p \to \varphi + anything,} \\ [24] \quad \mathrm{\pi^+p \to \varphi + anything} \end{cases}$$

have been studied at even lower energies ([32]), *i.e.* at 10 and 16 GeV/c (reaction [23]) for $\mathrm{K^-}$ momenta and 16 GeV/c (reaction [24]) π^+ momentum. Finally the following reactions ([33-35]):

(G)
$$\begin{cases} [25] \quad \mathrm{\gamma p \to \rho + anything,} \\ [26] \quad \mathrm{\gamma p \to \omega + anything,} \\ [27] \quad \mathrm{\gamma p \to \varphi + anything} \end{cases}$$

([30]) M. BARTH, C. DE CLERCQ, A. E. DE WOLF, J. J. DUMONT, D. P. JOHNSON, J. LE-
MONNE, P. PEETERS, R. CONTRI, H. DREVERMANN, L. GERDYUKOV, Y. GOLDSCHMIDT-
CLERMONT, G. HARIGEL, C. MILSTENE, J. P. PORTE, R. T. ROSS, P. THEOCHAROPOULOS,
C. CASO, F. FONTANELLI, R. MONGE, S. SQUARCIA, U. TREVISAN, J. F. BALAND,
J. BEAUFAYS, J. KESTEMAN, F. GRARD, P. A. VAN DER POEL, L. GATIGNON, W. KITTEL,
W. J. METZGER, D. J. SCHOTANUS, A. STERGIOU, R. T. VAN DE WALLE, Y. BELO-
KOPITOV, P. V. CHLIAPNIAKOV, A. FENUYK, V. KUBIC, S. LUGOVSKY, S. G. NIKITIN,
V. NIKOLAENKO, Y. PETROVIKH, V. RONJIN, O. TCHIKILEV and V. YARBA: *Z. Phys. C*,
7, 183 (1981).
([31]) M. BARTH, C. DE CLERCQ, E. DE WOLF, J. J. DUMONT, D. P. JOHNSON, J. LE-
MONNE, P. PEETERS, R. CONTRI, H. DREVERMANN, L. GERDYUKOV, Y. GOLDSCHMIDT-
CLERMONT, G. HARIGEL, J. JOENSUU, C. MILSTENE, J. P. PORTE, R. T. ROSS, M. SPY-
ROPOULOU-STASSINAKI, C. CASO, F. FONTANELLI, R. MONGE, S. SQUARCIA, U. TREVISAN,
J. F. BALAND, J. BEAUFAYS, F. GRARD, J. HANTON, P. A. VAN DER POEL, L. GATIGNON,
W. KITTEL, W. J. METZGER, D. J. SCHOTANUS, A. STERGIOU, R. T. VAN DE WALLE,
Y. BELOKOPITOV, B. BRIZGALOV, P. CHLIAPNIKOV, A. FENUYK, V. KUBIC, E. KRYT-
CHENKO, S. LUGOVSKY, V. NIKOLAENKO, J. PETROVIK, V. RONJIN, O. TCHIKILEV,
V. YARBA and V. ZJIGUNOV: *Z. Phys. C*, **7**, 89 (1981).
([32]) D. R. O. MORRISON: preprint CERN-EP/79-102 (1979), and references therein.
([33]) C. A. NELSON jr., E. N. MAY, J. ABRAMSON, D. E. ANDREWS, J. HARVEY, F. LOB-
KOWICZ, M. N. SINGER, E. H. THORNDIKE and M. E. NORDBERG jr.: *Phys. Rev. D*,
17, 647 (1978).
([34]) E. KOGAN, J. BALLAM, G. B. CHADWICK, K. C. MOFFEIT, P. SEYBOTH, I. O. SKIL-
LICORN, H. SPITZER, G. WOLF, H. H. BINGHAM, W. B. FRETTER, W. J. PODOLSKY,

140   M. BASILE, G. CARA ROMEO, L. CIFARELLI, A. CONTIN, G. D'ALÍ, ETC.

have been investigated at incident-photon momenta of 9.2 GeV/c for ρ and ω and 25.7 GeV/c for φ.

Reactions (A) and (B) are of the baryon-baryon type. Reactions (C) are of the antibaryon-baryon type. Reactions (D), (E) and (F) are of the meson-baryon type. In this latter class we include the (γp) processes (reactions (G)) because the « photon » is in fact a virtual vector meson and it allows us to extend our meson-baryon studies.

4'1. *The energy dependence of the « leading » quantity « L ».* – A study of the quantity L vs. the total c.m. energy of the colliding protons has been made, using Intersecting Storage Rings (ISR) data from $\sqrt{s} = 22$ GeV to $\sqrt{s} = 53$ GeV, for the inclusive neutron production [21]. No data are available for other hadrons in a wide energy range. This study shows that the quantity L, for the « neutron » case, scales. We have, however, pushed our study below the ISR energies, using Fermilab results at $\sqrt{s} = 13.7$ GeV c.m. energy. Comparing proton data at $\sqrt{s} = 13.7$ GeV with those at $\sqrt{s} = 62$ GeV, the maximum variation of the quantity L is found to be $\simeq 50\%$. However, this can be accounted for by strong variation of the « central » production of « protons » from Fermilab to ISR energies. This variation becomes very small in the ISR energy range. We will, therefore, assume that in the ISR energy range the quantity L scales for all types of hadrons.

4'2. *Study of the quantity « L » using different hadrons.* – As the main purpose of this section is to study the « leading » effect vs. different types of hadrons in the initial state, we will not include here the ISR results. The data obtained in (pp) collisions at ISR energies will be reported in subsect. 4'4. The main point to emphasize here is that the values of L for meson production at the ISR, *i.e.* in the highest-energy (pp) collisions, compare well with the values obtained at 100 GeV/c.

M. S. RABIN, A. H. ROSENFELD, G. SMADJA and Y. EISENBERG: *Nucl. Phys. B*, **122**, 383 (1977).
(35) D. ASTON, M. ATKINSON, R. BAILEY, A. H. BALL, B. BOUQUET, G. R. BROOKES, J. BRÖRING, P. J. BUSSEY, D. CLARKE, A. B. CLEGG, B. D'ALMAGNE, G. DE ROSNY, B. DIEKMANN, A. DONNACHIE, M. DRAPER, B. DREVILLON, I. P. DUERDOTH, J.-P. DUFEY, R. J. ELLISON, D. EZRA, P. FELLER, A. FERRER, P. J. FLYNN, F. FRIESE, W. GAL-BRAITH, R. GEORGE, S. D. M. GILL, M. GOLDBERG, S. GOODMAN, W. GRAVES, B. GROS-SETÊTE, P. G. HAMPSON, K. HEINLOTH, R. E. HUGHES-JONES, J. S. HUTTON, M. IB-BOTSON, M. JUNG, S. KATSANEVAS, M. A. R. KEMP, F. KOVACS, B. R. KUMAR, G. D. LAFFERTY, J. B. LANE, J.-M. LÉVY, V. LIEBENAU, J. LITT, G. LONDON, D. MERCER, J. V. MORRIS, K. MÜLLER, D. NEWTON, E. PAUL, P. PETROFF, Y. PONS, C. RAINE, F. RICHARD, R. RICHTER, J. H. C. ROBERTS, P. ROUDEAU, A. ROUGÉ, M. SENÉ, J. SIX, I. O. SKILLICORN, J. C. SLEEMAN, K. M. SMITH, C. STEINHAUER, K. M. STORR, D. TREILLE, CH. DE LA VAISSIÈRE, H. VIDEAU, I. VIDEAU, A. P. WAITE, A. WIJANGCO, W. WOJCIK, J.-P. WUTHRICK and T. P. YIOU: *Nucl. Phys. B*, **179**, 215 (1981).

4'2.1. Baryon-baryon collisions. As mentioned above, the baryon-baryon class is studied via the investigation of the multiparticle systems produced in proton-proton interactions at 100 GeV/c incident-proton momentum [25]

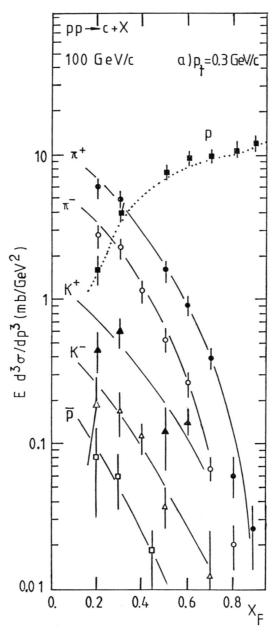

Fig. 2. – Inclusive production of hadrons in (pp) interactions. The invariant cross-section $E(\mathrm{d}^3\sigma/\mathrm{d}p^3)$ at 100 GeV/c as a function of x_F at fixed $p_\mathrm{T} = 0.3$ GeV/c, for p, $\pi^\pm$, $K^\pm$ and $\bar{\mathrm{p}}$ (ref. [25]).

142   M. BASILE, G. CARA ROMEO, L. CIFARELLI, A. CONTIN, G. D'ALÍ, ETC.

and at fixed $p_T = 0.3$ GeV/c. The data are shown in fig. 2, where the invariant $E(\mathrm{d}^3\sigma/\mathrm{d}p^3)$ inclusive cross-section for the production of p, $\pi^\pm$, $K^\pm$ and $\bar{p}$ as a function of x_F is reported. Notice that the nominal centre-of-mass energy is $\sqrt{s} = 13.6$ GeV. The x_F distributions for $\pi^\pm$, $K^\pm$ and $\bar{p}$ are peaked at $x_F = 0$, and decrease with increasing x_F values. The proton x_F distribution is very different: it is depressed at $x_F = 0$ and increases with increasing x_F. The value of the «leading» quantity L is $L_p \simeq 5$ for the proton; for all the other particles, L is an order of magnitude lower, $L < 0.5$, *i.e.* the proton is a «leading» particle in the final system of hadrons. The values of L are reported in fig. 3. With

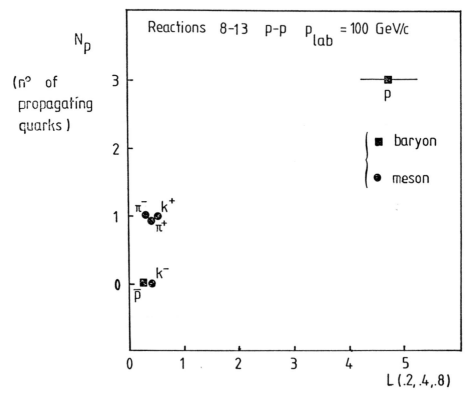

Fig. 3. – $L(0.2, 0.4, 0.8)$ for various final-state hadrons produced in pp collisions at $P_{lab} = 100$ GeV/c.

increasing transverse momentum, the x_R distribution is still very different from the produced particles' distributions, as can be seen from fig. 4 showing data from another spectrometer experiment [26]. In fig. 4 the proton inclusive cross-section is plotted as a function of x_F for different values of p_T: from 0.25 to 1.5 GeV/c. The experimental points at large x_R can be fitted at all $p_T \geqslant 0.75$ GeV/c by the unique form $(1 - x_R)^{0.5}$, as shown by the full lines in fig. 4. The «leading» effect is clearly present also in the large-p_T domain.

THE « LEADING »-PARTICLE EFFECT IN HADRON PHYSICS   **143**

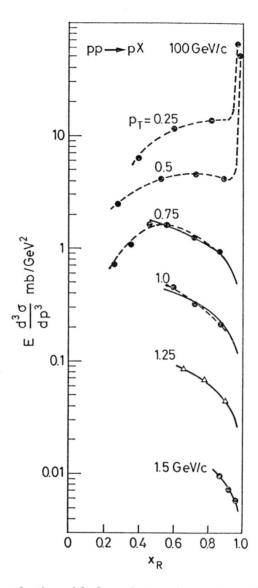

Fig. 4. – Inclusive production of hadrons in (pp) interactions. The invariant cross-section $E(\mathrm{d}^3\sigma/\mathrm{d}p^3)$ at 100 GeV/c as a function of x_R at various p_T from 0.25 to 1.5 GeV/c for protons (ref. ([26])). Open triangles at $p_T = 1.25$ GeV/c are data at 400 GeV/c. The dashed lines on the figure are hand-drawn to guide the eye through the data points; the full line is the shape of the $(1-x)^{0,5}$ functional behaviour.

4˙2.2. Baryon-antibaryon collisions. Note that in (pp) interactions the baryonic number is $B = 2$ and it must be conserved. In order to establish if the « leading » effect has something to do with the value of B, we have studied, at the same centre-of-mass energy, a reaction where $B = 0$, i.e. (p$\bar{\text{p}}$). Figure 5 shows data from a hydrogen bubble chamber experiment using

100 GeV/c antiprotons (^{27}). Here the invariant cross-sections $F(x)$ are plotted as a function of x_F for π^+, p and $\bar{p}$; backward π^- data have been averaged with the forward π^+ data. The antiproton x_F distribution has been determined, in

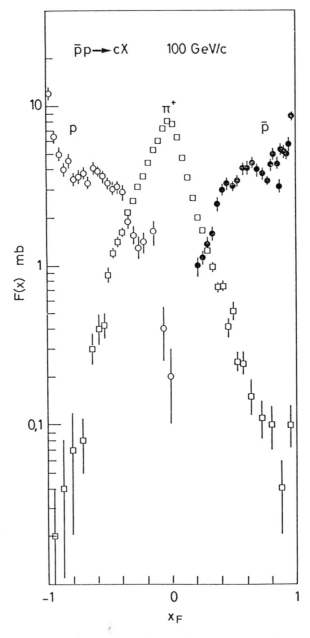

Fig. 5. – Inclusive production of hadrons in (p$\bar{p}$) interactions. Invariant cross-section integrated over p_T as a function of x_F for π^+, p and $\bar{p}$ as measured in 100 GeV/c $\bar{p}$ interactions in a hydrogen bubble chamber. (Data from ref. (^{27}).)

the x_F range $0.4 \leqslant x_F \leqslant 1.0$, using the $d\sigma/dp_{\text{lab}}$ cross-section from the same experiment; it is, within errors, the reflection of the proton x_F distribution at negative x_F values. The produced pion x_F distribution is peaked at $x_F = 0$ and falls off at large x_F in about the same way as in (pp) interactions. As can be seen from the data, both the protons in the $x_F < 0$ region and the antiprotons in the $x_F > 0$ region show a « leading » behaviour. The values of L are in fact again an order of magnitude larger for $\bar{p}$ and p than those of the produced particles.

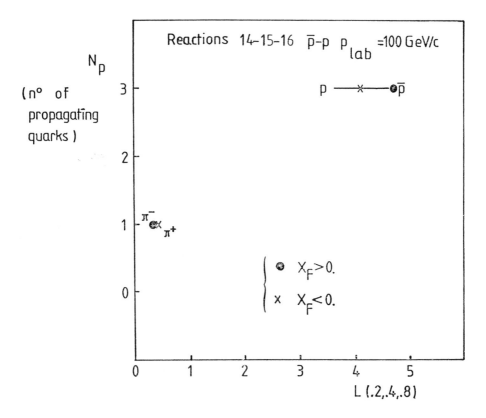

Fig. 6. – $L(0.2, 0.4, 0.8)$ for final-state hadrons produced in $\bar{p}$-p collisions at $P_{\text{lab}} = 100$ GeV/c.

This shows that a hadronic system which starts with two baryons behaves like the system which starts with a baryon-antibaryon pair. The « leading »-particle effect is present in the same way in both cases. The data are reported in fig. 6. For the produced pions the value of L is, as expected, $L_\pi < 0.5$.

4'2.3. Meson-baryon collisions. Let us now consider interactions where the incident hadron is a meson instead of a baryon. Figure 7a) shows,

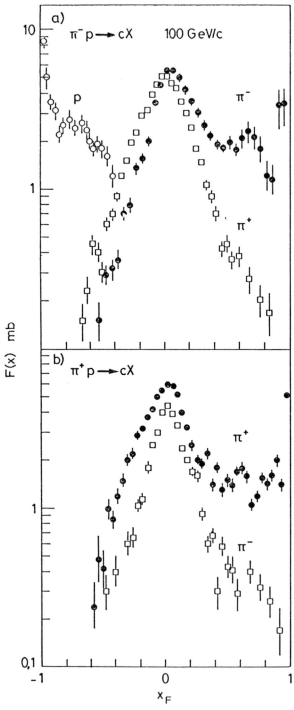

Fig. 7. – Inclusive production of hadrons in (πp) and (Kp) interactions. a) Invariant cross-section integrated over p_T as a function of x_F for π^+, π^- and p as measured in 100 GeV/c (π^-p) interactions in a hydrogen bubble chamber. (Data from ref. [28,29].) b) The same as for (a)) using 100 GeV/c (π^+p) interactions. (Data from ref. [28].)

at 100 GeV/c π^- beam momentum, the π^+, π^- and proton inclusive invariant x_F distributions integrated over p_T in (πp) interactions. The data are from the 30 inch hydrogen bubble chamber/wide-gap spark chamber hybrid system at Fermilab [28,29]. Here, both the proton and the π^- show a «leading» effect.

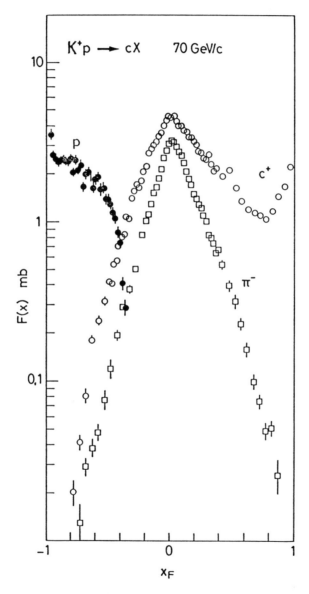

Fig. 7. – Inclusive production of hadrons in (πp) and (Kp) interactions. c) Invariant cross-section at fixed $p_T = 0.3$ GeV/c as a function of x_F for inclusive production in (K$^+$p) interactions at 70 GeV/c in BEBC. (Data from ref. [30,31].) Shown is the x_F distribution of positively charged mesons c$^+$ (slow protons subtracted) of π^- and of the target protons.

The values of L for p and π^- are $L_p \geqslant 3$ and $L_{\pi^-} \simeq 2.5$, respectively. These results are shown in fig. 8. The only difference between the « leading » baryon and the « leading » meson is that the latter is also produced abundantly at $x_F = 0$. Notice that the value of L for π^+ is much smaller than the π^- value, as expected for a particle which is produced without any « leading » effect.

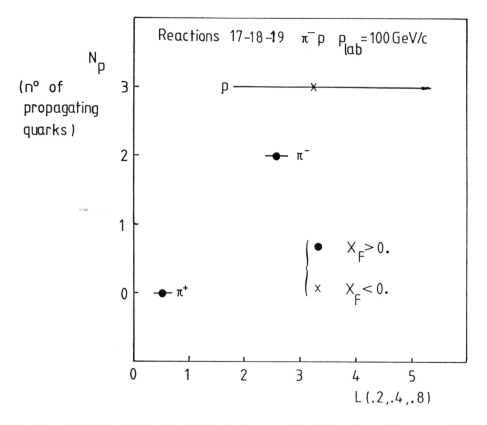

Fig. 8. – $L(0.2, 0.4, 0.8)$ for final-state hadrons produced in π^-p at $P_{\text{lab}} = 100$ GeV/c.

Figure 7b) shows π^+ and π^- x_F distributions in $(\pi^+\text{p})$ interactions at 100 GeV/c. It can be clearly seen that the π^+ x_F distribution closely resembles the π^- x_F distribution of fig. 7a). Both sets of data come from the same experiment ([28]) and can be directly compared: they show the same « leading » effect for π^+ and π^-. The value of L_{π^+} is in fact the same as the L_{π^-} mentioned above.

In fig. 7c) the (K^+p) data obtained at 70 GeV/c by the BEBC Collaboration ([30,31]) are presented. Here the K^+ are not resolved from the π^+. The x_F distribution of the positively charged particle c^+ must, therefore, be compared with that of the π^- and of the target proton. Although not resolved from π^+, the K^+ « leading » effect for $x_F > 0$ is clear. Also the target proton « leading » effect for $x_F < 0$ is equally clear. The values of L_{c^+} and of L_p are shown in fig. 9.

Finally we turn to vector mesons. These are not available in hadron beams, but—because of their J^{PC} quantum numbers—they can be produced in photoproduction experiments. The reaction (γp) can, therefore, be regarded as a beam

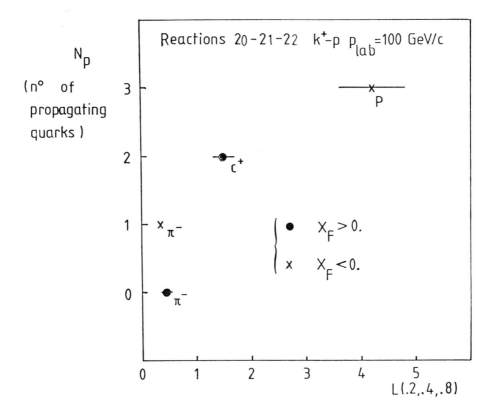

Fig. 9. – $L(0.2, 0.4, 0.8)$ for final-state hadrons produced in K$^+$p at $P_{lab} = 100$ GeV/c.

of ρ, ω, φ and other vector mesons interacting with the proton. Figure 10 shows the x distribution of the vector mesons inclusively produced in the reaction

$$ (6) \qquad\qquad \gamma + p \to V + X . $$

The data are very limited; (ρ and ω) photoproductions were studied at 9.2 GeV, while φ photoproduction was studied at 25.7 GeV energy of incident photon. The data are from various photoproduction experiments [33-35]. The « leading »-vector-meson effect is clearly visible in the almost flat x distribution in the beam x-region $x_F > 0$. The values of L are shown in fig. 11. Notice the anomalously large leading behaviour of the ω vector meson.

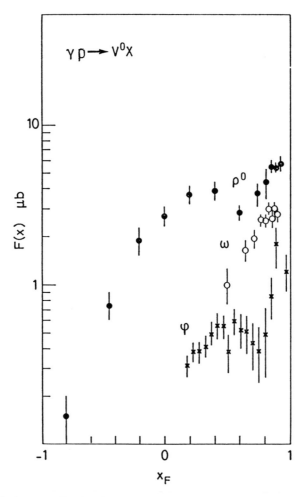

Fig. 10. – Inclusive x_F distribution of vector mesons in (γp) interactions: ρ meson (ref. ([33])) and ω meson (ref. ([34])) at γ energy of 9.2 GeV. The φ-meson distribution (plotted with arbitrary normalization from ref. ([35])) is for a γ energy of 25.70 GeV.

4`3. *The « leading » effect when the incident particle does not change its nature.* – From this review it is evident that the « leading » effect is present in all hadron-hadron interactions. This effect should not be mistaken for the diffractive peak at x_F (x_R) $\simeq 1$. The « leading »-hadron effect shows up in a very important x-range, typically from 0.2 to 0.8. This range is enough far away from the « central » production, $x \simeq 0$, and from the « peripherality » region, $x \simeq 1$. The initial-state hadron shows up in the final state as a « leading » particle, with a nearly flat fractional energy or momentum distribution (x_F, x_R). Notice that the diffractive peak disappears for $p_T \gtrsim 0.7$ GeV/c, but the « leading » effect remains.

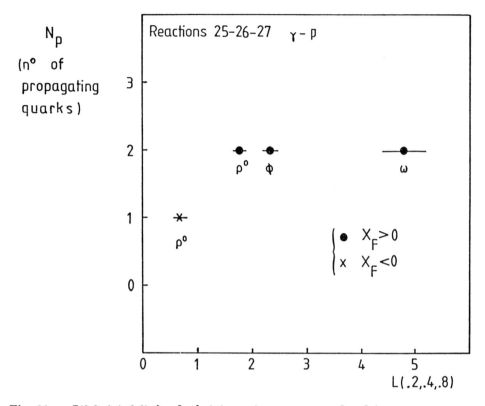

Fig. 11. – $L(0.2, 0.4, 0.8)$ for final-state vector mesons produced in γp.

The «leading» effect seems naturally connected with the «quantum number flow» from the initial state to the particle in the final state. These quantum numbers—flavour, J^{PC} and colour—are carried by the constituents which make up the initial-state hadron. The particle which carries more constituents of the initial state will have a sizable fraction of the four-momentum of the initial state, and will hence appear to be «leading».

There is, however, a difference between mesons and baryons. Mesons show a «leading» effect and a sizable central production. In fact the ratio

$$\frac{(\mathrm{d}\sigma/\mathrm{d}x)_{x=0}}{(\mathrm{d}\sigma/\mathrm{d}x)_{x\simeq0.6}}$$

tends, for mesons, to be bigger than 1, whereas for baryons it is of the order of or smaller than 1. This can be naively explained by the simple fact that it is easier to produce a meson-antimeson than a baryon-antibaryon pair. In fact, it is easier to recombine a quark-antiquark pair in order to obtain a meson than three quarks (antiquarks) to obtain a baryon (antibaryon). In fig. 12a) we show, for comparison, the values of $F(x)$ vs. x_F of different hadron beams

M. BASILE, G. CARA ROMEO, L. CIFARELLI, A. CONTIN, G. D'ALÍ, ETC.

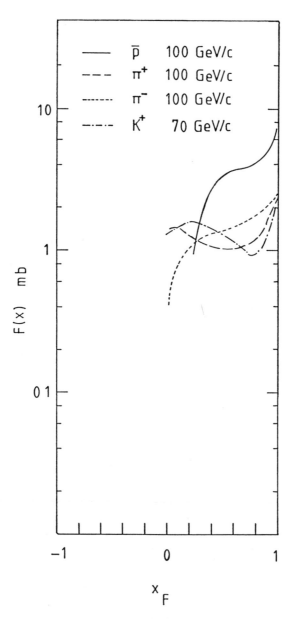

Fig. 12. – a) x_F distribution in the beam x_F-region ($x_F > 0$) for $\pi^\pm$, K^+ and $\bar{p}$ incident particles on the target protons.

on a proton target. The fits are superimposed on the same figure so as to compare the shape in the $x_F > 0$ region. The curves are obtained from the data of fig. 5 and 7. For π^-, π^+ and K^+ mesons the contributions from centrally produced mesons have been subtracted using, respectively, the π^+, π^- and K^- distributions. For these particles in fact the « leading » effect in not present, and they provide a good check for the « standard » production mechanism. We have computed the value of the « leading » quantity « L » for the different beam particles excluding, as usual, the region below $x_F = 0.20$ (dominated

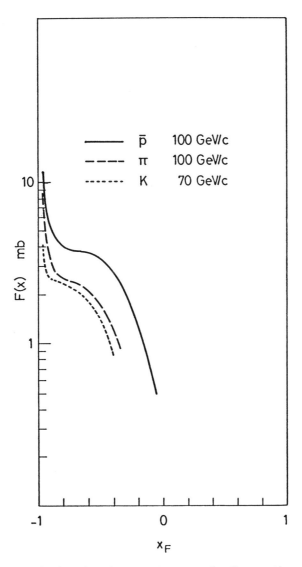

Fig. 12. – b) x_F distribution for the target proton in the reactions (πp), (Kp), ($\bar{p}$p).

by the central production) and above $x_F = 0.80$ (to eliminate diffractive production). The results are given in the first five rows of table II.

Figure 12b) shows a comparison of the « leading » proton in the target x_F-region in reactions initiated by different beam particles. The corresponding « L » values, computed in the same x_F-region as discussed above, are reported

154   M. BASILE, G. CARA ROMEO, L. CIFARELLI, A. CONTIN, G. D'ALÍ, ETC.

TABLE II. – *Value of the «leading» quantity «L» measured for various types of hadron beams, i.e.* π^+, π^-, K^+, p, $\bar{p}$.

Reaction	x_F region	«L»
$\pi^+ p \to \pi^+ +$ any	> 0	2.5 ± 0.3
$\pi^- p \to \pi^- +$ any	> 0	2.5 ± 0.3
$K^+ p \to K^+ +$ any	> 0	1.5 ± 0.2
$pp \to p +$ any	> 0	5 ± 0.5
$\bar{p}p \to \bar{p} +$ any	> 0	4.5 ± 0.5
$\bar{p}p \to p +$ any	< 0	4 ± 0.5
$K^+ p \to p +$ any	< 0	4 ± 0.5
$\pi^- p \to p +$ any	< 0	3 ± 0.4

in the last three rows of table II. These data show that, no matter the nature of the incident particle, the proton keeps its «leading» effect, as expected.

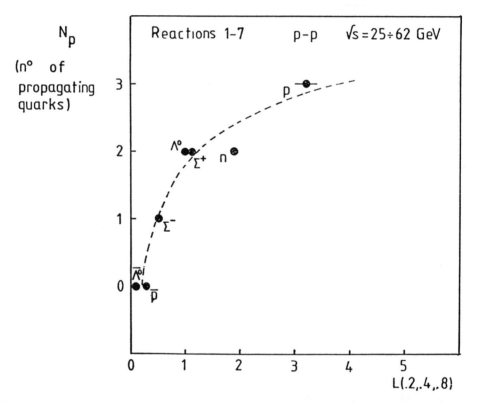

Fig. 13. – The leading quantity $L(0.2, 0.4, 0.8)$ for various final-state hadrons in pp collisions at ISR energies $((25 \div 62) \, \text{GeV})$ is plotted *vs.* the number of propagating quarks from the incoming into the final-state hadrons. The dashed line is obtained by using a parametrization of the single-particle inclusive cross-section, as described in the text.

Finally, for the (pp) case, the ISR results (which will be fully reported in the next section) show a clear «leading» effect.

So far, the «leading» effect has been studied when the incident particle goes from the initial state to the final state without changing its nature: a proton remains a proton, a π^+ remains a π^+, etc. In the next subsection we will consider the case where the incident particle changes its nature.

4˙4. *The « leading » effect when the incident particle changes its nature.* – What about the case of an incident hadron which transforms itself into another type of hadron? For example, a proton which becomes a Λ^0? There is definite evidence of the following fact: the «leading» phenomenon follows the rule that the « more similarity » there is (for example, in the sense of the quark content) between the hadronic particle observed in the final-state and the initial-state hadron, the more the «leading» effect is present. As mentioned in sect. **2**, the quantity « L » scales in the ISR energy range. Therefore, the cleanest way of studying the « leading » effect when a hadron changes its nature, is to measure the quantity « L » for the various baryons produced in (pp) interactions at the ISR ([21-24]). As discussed at the beginning of this section, the reactions investigated were

$$
\begin{cases}
[1] & \text{pp} \to \text{p}+\text{anything,} \\[4pt]
[2] & \text{pp} \to \text{n}+\text{anything,} \\[4pt]
[3] & \text{pp} \to \Lambda+\text{anything,} \\[4pt]
[4] & \text{pp} \to \Sigma^+ +\text{anything,} \\[4pt]
[5] & \text{pp} \to \Sigma^- +\text{anything,} \\[4pt]
[6] & \text{pp} \to \bar{\text{p}}+\text{anything,} \\[4pt]
[7] & \text{pp} \to \overline{\Lambda}+\text{anything.}
\end{cases}
$$

Figure 13 shows the values of L for different hadrons produced in proton-proton collisions in the ISR energy range from 25 to 62 GeV. The dashed line is obtained using for L the functional form

$$F(x) = (1 - x)^{2n-1},$$

where n is the number of quarks that need to be changed in order to give rise to the final wanted hadron ([36-38]). This allows a comparison of our « leading »

([36]) D. DENEGRI, C. COCHET, M. A. JABIOL, C. LEWIN, L. MOSCA, M. L. FACCINI-TURLUER, D. VILANOVA, L. BECKER, U. GENSCH, M. BARTH, E. A. DE WOLF, M. GYSEN, M. VAN IMMERSEEL, C. POIRET, R. WINDMOLDERS, I. V. AJINENKO, YU. ARESTOV, A. A. BOROVIKOV, P. V. CHLIAPNIKOV, V. V. KNIAZEV, E. A. KOZLOVSKY, M. N.

quantity L with old parametrization methods where the $F(x)$ was parametrized as given above. The agreement is excellent. In table III the values of L shown in fig. 13 are reported, together with the number of «propagating» quarks corresponding to the final-state hadron studied. In fig. 13 the final states have been grouped according to the number of quarks propagating from the initial to the final state. For example, if we start with a proton, whose quark composition is (udu), and look at the Λ^0 whose quark composition is (uds), the propagating quarks are two (ud).

TABLE III. – ISR (pp) *processes*.

Initial state: proton (udu)			
Final-state particle	Quark composition	Propagating quarks	L
p	(udu)	(udu)	3.2 ± 0.2
n	(udd)	(ud)	1.92 ± 0.05
Λ^0	(uds)	(ud)	1.02 ± 0.10
Σ^+	(usu)	(uu)	1.15 ± 0.17
Σ^-	(dds)	(d)	0.53 ± 0.15
$\bar{p}$	$(\bar{u}\bar{d}\bar{u})$	nothing	0.30 ± 0.05
$\bar{\Lambda}^0$	$(\bar{u}\bar{s}\bar{d})$	nothing	0.10 ± 0.02

The results shown in fig. 3, 6, 8 and 9, 11 and 13, have a common trend, characterized by the number of propagating quarks: the value of L is the same when the number of propagating quarks is the same.

Recent studies on heavy-flavoured baryon production in (pp) collisions [10-14] clearly show the leading effect to be present also in Λ_c^+ and Λ_b^0 production.

The presence of the «leading» effect, connected to the number of propagating quarks, is clearly shown also by the investigation of the reactions

$$\begin{cases} [23] & \mathrm{K^- p} \to \varphi + \text{anything}, \\ [24] & \pi^+ \mathrm{p} \to \varphi + \text{anything}. \end{cases}$$

UKHANOV, E. V. VLASOV, H. DIBON and M. MARKYTAN: *Phys. Lett. B*, **98**, 127 (1981), and references therein.

[37] D. CUTTS, R. S. DULUDE, R. E. LANOU jr., J. T. MASSIMO, R. MEUNIER, A. E. BRENNER, D. C. CAREY, J. E. ELIAS, P. H. GARBINCIUS, G. MIKENBERG, V. A. POLYCHRONAKOS, M. D. CHIARATIA, C. DEMARZO, C. FAVUZZI, G. GERMINARIO, L. GUERRIERO, P. LaVOPA, G. MAGGI, F. POSA, G. SELVAGGI, P. SPINELLI, F. WALDNER, W. AITKENHEAD, D. S. BARTON, G. W. BRANDENBURG, W. BUSZA, T. DOBROWOLSKI, J. I. FRIEDMAN, H. W. KENDALL, T. LYONS, B. NELSON, L. ROSENSON, W. TOY, R. VERDIER and L. VOTTA: *Phys. Rev. Lett.*, **43**, 319 (1979).

[38] J. S. GUNION: *Phys. Lett. B*, **88**, 150 (1979).

For reaction [23] the hadron in the final state, $\varphi(s\bar{s})$, contains one of the quarks which make up the hadron in the initial state, $K^-(s\bar{u})$. Only one quark propagates in this case. For reaction [24] there is no quark propagation when going from a $\pi^+(u\bar{d})$ to a φ-meson $(s\bar{s})$. The data [32] of fig. 14 clearly show a « leading » effect for the φ produced in reaction [23]. The value of L is $L_\varphi \simeq 4.5$. The «leading» effect is not present in φ production from reaction [24], as expected. The value of L is $L_\varphi \simeq 1$. It should be noticed that the value of L increases with decresing energy of the incident hadron. This has a straightforward interpretation. At lower energy the « central » production is depressed. This produces a smaller denominator in our definition of L. The purpose of the analysis of reactions [23] and [24] was to see if even at this extreme low energy there was a clear difference in the L values for different numbers of propagating quarks.

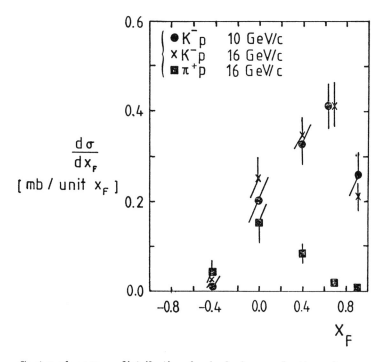

Fig. 14. – Centre-of-mass x_F distribution for inclusive production of φ-mesons in K^-p and π^+p interactions (ref. [32]).

5. – The « leading » hadron in lepton-hadron collisions.

In lepton-hadron reactions, *i.e.* (ep), (μp) and (νp) deep inelastic scattering, the energy available for hadron production is not the total centre-of-mass energy, but only the energy lost by the lepton. According to the standard formalism, we call $(E - E')$ the energy lost by the lepton, Q^2 the square of the four-mo-

158   M. BASILE, G. CARA ROMEO, L. CIFARELLI, A. CONTIN, G. D'ALÍ, ETC.

mentum transfer (changed in sign) and W the invariant mass of the final-state hadronic system recoiling against the lepton.

Experimental results on the production of hadrons from (μp) deep inelastic scattering experiments and from high-energy (νp) interactions are at present available—although not always with good particle identification. For our study we have used data on hadron production in (νp) and ($\bar{\nu}$p) charged-current events [39,40]. The data refer to exposures of the Fermilab 15 ft bubble chamber to broad-band ν and $\bar{\nu}$ beams. The x_F distributions of positive (h$^+$) and negative (h$^-$) hadrons are measured both in the forward and in the backward hemispheres. In the h$^+$ sample, protons can be identified, either by ionization or by kinematical fitting to the final states, and their x_F distribution can be derived. The most significant data refer to the reaction $\bar{\nu}$p $\to \mu^+ + p+$anything, at $W \simeq 5$ GeV [40]. These are reported in fig. 15. The h$^+$ distribution, as a function of x_F, shows a flat behaviour in the recoil hemisphere ($x_F < 0$), which is not present in the h$^-$ case. Such an effect can be explained in terms of the presence of « leading » protons. In fact, for protons identified in the h$^+$ sample, the x_F distribution shows a plateau in the $x_F < 0$ region, thus providing direct evidence of the « leading »-proton effect in ($\bar{\nu}$p) interactions. The value of L, derived from the data of fig. 15, is $L_p = 3.2 \pm 0.5$. Similar features are shown by (νp) charged-current events, as described in ref. [39].

The « leading »-baryon effect is also present in the inclusive x_F distribution of the Λ^0 baryon produced both in electromagnetic and weak interactions. As we have seen in subsect. 4˙4, the Λ^0 is as « leading » as a baryon should be, when there are two propagating quarks. Figure 16 is a compilation [41] of production cross-sections of the Λ^0 in (ep), (νp) and ($\bar{\nu}$p) interactions: the x_F distributions have been normalized so as to compare their shape. The Λ^0 exhibits a clear « leading » effect in electromagnetic (ep), and weak (νp)

[39] J. BELL, C. T. COFFIN, R. N. DIAMOND, H. T. FRENCH, W. C. LOUIS, B. P. ROE, R. T. ROSS, A. A. SEIDL, J. C. VANDER VELDE, E. WANG, J. P. BERGE, D. V. BOGERT, F. A. DiBIANCA, R. ENDORF, R. HANFT, C. KOCHOWSKI, J. A. MALKO, G. I. MOFFATT, F. A. NEZRICK, W. G. SCOTT, W. SMART, R. J. CENCE, F. A. HARRIS, M. JONES, M. W. PETERS, V. Z. PETERSON, V. J. STENGER, G. R. LYNCH, J. P. MARRINER and M. L. STEVENSON: *Phys. Rev. D*, **19**, 1 (1979).

[40] M. DERRICK, P. GREGORY, F. LoPINTO, B. MUSGRAVE, J. SCHLERETH, P. SCHREINER, R. SINGER, S. J. BARISH, R. BROCK, A. ENGLER, T. KIKUCHI, R. W. KRAEMER, F. MESSING, B. J. STACEY, M. TABAK, V. E. BARNES, T. S. CARMAN, D. D. CARMONY, E. FERNANDEZ, A. F. GARFINKEL and A. T. LAASANEN: *Phys. Rev. D*, **24**, 1071 (1981).

[41] V. AMMOSOV, A. AMRAKHOV, A. DENISOV, P. ERMOLOV, V. GAPIENKO, V. KLYU-KHIN, V. KORESHEV, P. PITUKHIN, V. SIROTENKO, E. SLOBODYUK, V. ZAETZ, J. P. BERGE, D. BOGERT, R. HANFT, J. MALKO, G. HARIGEL, G. MOFFAT, F. NEZRICK, J. WOLFSON, V. EFREMENKO, A. FEDOTOV, P. GORICHEV, V. KAFTANOV, G. KLIGER, V. KOLGANOV, S. KRUTCHININ, M. KUBANTSEV, I. MAKHLYUEVA, V. SHEKELJAN, V. SHEVCHENKO, J. BELL, C. T. COFFIN, W. LOUIS, B. P. ROE, R. T. ROSS, D. SINCLAIR and E. WANG: *Nucl. Phys. B*, **162**, 205 (1980), and references therein.

THE « LEADING »-PARTICLE EFFECT IN HADRON PHYSICS $\qquad$ **159**

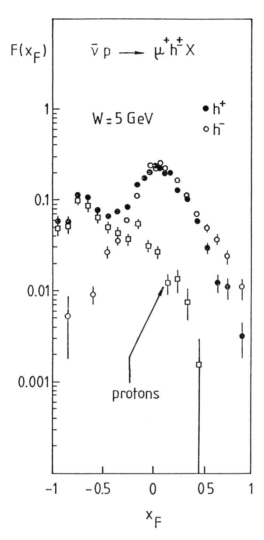

Fig. 15. – Inclusive production of hadrons in ($\bar{\nu}$p) charged-current interactions. The centre-of-mass x_F distributions for positive-charge particles (h$^+$), for negative-charge particles (h$^-$), and for identified protons (ref. (^{40})).

and ($\bar{\nu}$p), interactions. The best data for measuring L_{Λ^0} are those for the reaction

[28] $\qquad \bar{\nu}\mathcal{N} \to \Lambda^0 + \text{anything}$,

referring to Fermilab 15 ft bubble chamber data at $W \simeq 4.5$ GeV. However, L_{Λ^0} can also be determined from data taken in a streamer chamber experiment at Cornell for the process

[29] $\qquad \text{ep} \to \Lambda^0 + \text{anything}$,

160   M. BASILE, G. CARA ROMEO, L. CIFARELLI, A. CONTIN, G. D'ALÍ, ETC.

at the energy $W \simeq 3$ GeV. The values of L_{Λ^0} are reported in fig. 17. As mentioned above, the data show that the «leading» effect in Λ^0 production, first observed in purely hadronic interactions, is also present in electromagnetic and weak interactions. In fact, the Λ^0 production corresponds to a

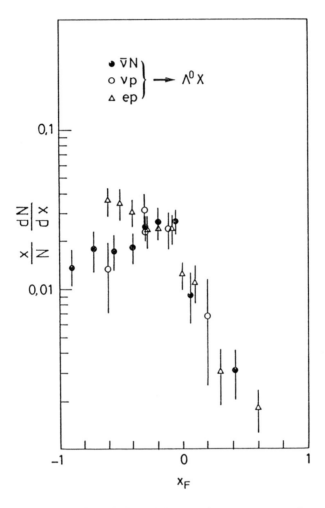

Fig. 16. – Inclusive production of the Λ^0 baryon in $(\bar{\nu}N)$, (νp) and (ep) interactions. Data from ref. ([41]) are normalized to the same cross-section.

proton which transforms into a Λ^0 via the propagation of two quarks. The value of L_{Λ^0} is as expected when the number of propagating quarks is two.

It is interesting to remark that also the energy dependence of L_{Λ^0} follows the trend already observed for the proton case (see subsect. 4'1). At ISR energies, the value of L scales. At Fermilab energies, down to the energy of reaction [29], the behaviour of L can be summarized as follows: the lower is the energy available, the higher is the value of L.

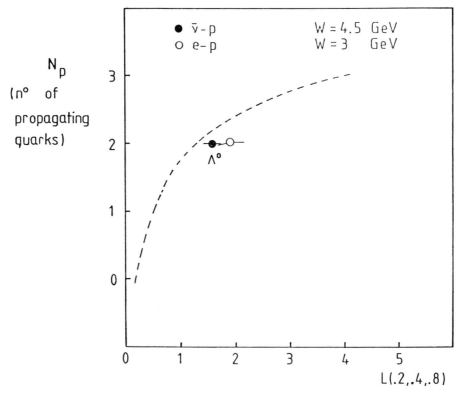

Fig. 17. – $L(0.2, 0.4, 0.8)$ for Λ^0 productions in ($\bar{\nu}p$) and (ep) reactions. The dashed line is the same as that described for fig. 13.

6. – Conclusions.

Thanks to the use of the « leading » effect it has been possible to compare ([1-9]) the production of multihadron systems in (pp) interactions and (e+e−) annihilation. So far, the main trend had been that only high–transverse-momentum hadronic interactions could be a source of comparison with (e+e−) and deep inelastic scattering (DIS) in lepton-hadron physics. The use of the « leading »-hadron effect has brought the vast amount of low-p_T physics into the very interesting domain of comparison with (e+e−) and DIS.

The study reported in the present paper shows that, when a hadron interacts with another particle hadronically, electromagnetically, or weakly, a very distinct feature emerges: the hadron, which is present in the initial state, keeps a privileged role in the energy sharing with the other hadrons produced. This distinctive feature is measured by the « leading » quantity « L ». It is found that the value of this quantity is $L \sim 3$, when the number of propagating quarks is 3. If the number of propagating quarks is 2, the value of L is $L \simeq 1.5$. If only one quark propagates from the initial to the final state, the value of L is $L \simeq 0.5$. When there is no quark propagation from the initial to the final state, the particle produced has a value of the leading quantity $L < 0.5$. This value of L holds true, also, when a hadron is procuced in (e⁺e⁻) annihilation. Here no hadron has a privileged energy-sharing, *i.e.* there is no « leading » hadron. In fact, in (e⁺e⁻) annihilation, in the initial state, there is a certain amount of energy E, to be used for the production of hadrons, but no hadron is present in the initial state.

For the « leading » effect to show up, one needs an incoming hadron carrying good quantum numbers: colour, flavour, J^{PC}, etc. It is this quantum number « flow » which generates the « leading » phenomenon. Even when the incident particle changes its nature (for example, a proton becomes a neutron, or a Λ^0, or a Λ_c^+), *i.e.* when the initial-state quantum numbers are not fully carried by the final-state particle, the « leading » effect is still present.

The « leading » phenomenon, therefore, remains the cleanest effect, even in processes in which quantum numbers become mixed and interchanged. Here we have a field of physics, the hadron constituents' dynamics, which is far from being understood and is, therefore, difficult to take into account. In spite of all this, the « leading » effect shows up as a clear signature of simplifying unity.

For example, in deep inelastic lepton-hadron scattering, the target hadron keeps a large fraction of the available energy. This is currently attributed to the gluons, but it is simply a « leading »-baryon effect, whose significance in terms of gluons is a matter of theoretical speculation, as the quantum number « flow » mentioned above.

The use of the « leading » effect should be extended to the study of all hadronic phenomena. Its application to the analysis of multiparticle hadronic states produced in high-energy interactions will draw a more clear and more uniform picture in the study of processes induced by different particles at different energies.

● RIASSUNTO

Si presenta una rassegna mondiale dell'effetto di particella « leading » in fisica adronica. Processi del tipo barione-barione, barione-antibarione, mesone-barione e fotone-barione sono studiati e confrontati con la produzione di adroni da uno stato iniziale che non contiene adroni. In quest'ultimo tipo di processi non è presente l'effetto « leading ». Si dimostra che l'effetto di particella « leading » è sicuramente presente anche nelle interazioni iniziate da neutrini con adroni nello stato finale. Si discute infine l'importanza dell'effetto « leading » nella fisica adronica.

Эффект «лидирующей» частицы в адронной физике.

Резюме (*). — Проводится анализ эффекта «лидирующей» частицы в адронной физике. Исследуются барион-барионные, барион-антибарионные, мезон-барионные и протон-барионные процессы. Проводится сравнение с процессом образования адрона из начального состояния, которое не содержит адрона. В этом последнем случае эффект «лидирующей» частицы отсутствует. Показывается, что эффект «лидирующей» частицы имеет место в адронных взаимодействиях, инициированных нейтрино. Обсуждается важность эффекта «лидирующей» частицы в адронной физике.

(*) *Переведено редакцией.*

CERN
SERVICE D'INFORMATION
SCIENTIFIQUE

M. BASILE, *et al.*
21 Novembre 1981
Il Nuovo Cimento
Serie, 11, Vol. **66** A, pag. 129-163

IL NUOVO CIMENTO   VOL. 79 A, N. 1   1 Gennaio 1984

Universality Features in (pp), (e⁺e⁻) and Deep-Inelastic-Scattering Processes.

M. Basile, G. Bonvicini, G. Cara Romeo, L. Cifarelli, A. Contin
M. Curatolo, G. D'Alí, C. Del Papa, B. Esposito, P. Giusti, T. Massam
R. Nania, F. Palmonari, G. Sartorelli, G. Susinno, L. Votano
and A. Zichichi

CERN - Geneva, Switzerland
Istituto di Fisica dell'Università - Bologna, Italia
Istituto Nazionale di Fisica Nucleare, Laboratori Nazionali di Frascati - Frascati, Italia
Istituto Nazionale di Fisica Nucleare - Sezione di Bologna, Italia

(ricevuto il 29 Luglio 1983)

Summary. — The use of the correct variables in (pp), (e⁺e⁻), and deep-inelastic scattering (DIS) processes allows universality features to be established in these—so far considered—different ways of producing multihadronic states.

PACS. 25.30. – Lepton-induced reactions and scattering.
PACS. 25.40. – Nuclear-induced reactions and scattering.

1. – Introduction.

Multihadronic final states can be produced in purely hadronic interactions such as (pp), in purely electromagnetic interactions such as (e⁺e⁻), and in lepton-hadron deep-inelastic-scattering (DIS) processes, which can be either weak (such as νp) or electromagnetic (such as μp).

The purpose of this paper is to review the main points needed in order to establish a common basis for a comparison between these various ways of producing multihadronic states. The interest in this study is twofold:

i) So far, these three ways of producing multiparticle systems have been considered to be basically different. Our study shows that, before reaching

1

such a conclusion, it is imperative to use the correct variables in describing the three processes: (pp), (e⁺e⁻) and DIS. In fact, the use of the correct variables allows universal features to be revealed in the multihadronic final states produced in (pp), (e⁺e⁻) and DIS.

ii) Once these—so far considered—different ways of producing multihadronic final states are brought within the correct framework, the comparison can be made at a deeper level, and basic differences can thus be studied, if they exist. These differences must be at a level which is below our present one, where we have established striking common features.

We will see that (pp) interactions can be studied *à la* (e⁺e⁻) and *à la* DIS. However, the DIS way turns out to be incorrect when comparing DIS data with those obtained in (e⁺e⁻). The equivalence found between (pp) and (e⁺e⁻) data allows the « differences » reported between DIS and (e⁺e⁻) data to be understood. These differences are reproduced if (pp) data are analysed *à la* DIS, and disappear when the correct variables are used.

In sect. **2** we introduce the correct variables. In sect. **3** we report all results obtained so far when comparing (pp), (e⁺e⁻) and DIS. In sect. **4** we extrapolate our findings to the multihadronic systems to be studied at collider energies, such as at the CERN (pp̄) machine. Section **5** gives the conclusions.

Notice that the universality features discovered are a zero-parameter fit to the various properties of the multihadronic systems produced in (pp), (e⁺e⁻), and DIS.

2. – The identification of the correct variables.

The identification of the correct variables for describing hadron production in (pp) interactions, (e⁺e⁻) annihilation and DIS processes is the basic starting point for putting these three ways of producing multiparticle hadronic systems on an equal footing. In this section we show how this can be done.

2'1. – e⁺e⁻ *annihilation* is illustrated in fig. 1, where q_1^{inc} and q_2^{inc} are the four-momenta of the incident electron e⁻ and positron e⁺; q^{h} is the four-momentum of a hadron produced in the final state, whose total energy is

$$(1) \qquad (\sqrt{s})_{e^+e^-} = \sqrt{(q_1^{\text{inc}} + q_2^{\text{inc}})^2} = 2E_{\text{beam}}$$

(when the colliding beams have the same energy).

As we will see later,

$$(2) \qquad q_1^{\text{inc}} = q_1^{\text{had}}, \qquad q_2^{\text{inc}} = q_2^{\text{had}},$$

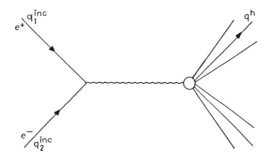

Fig. 1. – Schematic diagram for the (e+e−) annihilation.

where $q_{1,2}^{had}$ are the four-momenta available in a (pp) collision for the production of a final state with total hadronic energy

$$(3) \qquad \sqrt{(q_1^{had} + q_2^{had})^2} = \sqrt{(q_{tot}^{had})^2}.$$

It is this quantity $\sqrt{(q_{tot}^{had})^2}$ which should be used in the comparison with (e+e−) annihilation, and therefore with

$$(4) \qquad (\sqrt{s})_{e+e-}.$$

This means that

$$(5) \qquad (\sqrt{s})_{e+e-} = \sqrt{(q_{tot}^{had})^2}.$$

Moreover, the fractional energy of a hadron produced in the final state of an (e+e−) annihilation is given by

$$(6) \qquad (x)_{e+e-} = 2\, \frac{q^h \cdot q_{tot}^{had}}{q_{tot}^{had} \cdot q_{tot}^{had}} = 2\, \frac{E^h}{(\sqrt{s})_{e+e-}},$$

where the dots indicate the scalar product and E^h is the energy of the hadron « h » measured in the (e+e−) c.m. system, where that the four-momentum q_{tot}^{had} has no spacelike part:

$$(7) \qquad q_{tot}^{had} \equiv [i\mathbf{0}; (\sqrt{s})_{e+e-}].$$

2`2. – DIS *processes* are illustrated in fig. 2, where q_1^{inc} and $q_1^{leading}$ are the four-momenta of the initial- and final-state leptons, respectively; q_2^{inc} is the four-

4

momentum of the target nucleon; q_1^{had} is the four-momentum transferred from the leptonic to the hadronic vertex, whose timelike component in the laboratory reference system is coincident with the invariant quantity, usually indicated as ν:

$$(8) \qquad q_1^{\text{had}} \equiv (i\boldsymbol{p}_1^{\text{had}}; \nu \equiv E_1^{\text{had}}) \,.$$

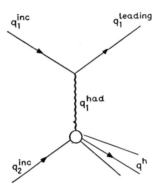

Fig. 2. – Schematic diagram for the DIS processes.

Notice that, in order to easily identify the equivalent variables in (pp) interactions, we have introduced a notation in terms of E_1^{had} and $\boldsymbol{p}_1^{\text{had}}$.

A basic quantity in DIS is the total hadronic mass

$$(9) \qquad (W^2)_{\text{DIS}} = (q_1^{\text{had}} + q_2^{\text{inc}})^2 \,,$$

and the fractional energy is

$$(10) \qquad (z)_{\text{DIS}} = \frac{q^{\text{h}} \cdot q_2^{\text{inc}}}{q_1^{\text{had}} \cdot q_2^{\text{inc}}} \,,$$

where again the dots between the four-momenta indicate their scalar product.

$2'3.$ – (pp) *interactions* are illustrated in fig. 3, where $q_{1,2}^{\text{inc}}$ are the four-momenta of the two incident protons, $q_{1,2}^{\text{leading}}$ are the four-momenta of the two leading protons, $q_{1,2}^{\text{had}}$ are the spacelike four-momenta emitted by the two proton vertices, q^{h} is the four-momentum of a hadron produced in the final state.

Now, attention! A (pp) collision can be analysed in such a way as to produce the key quantities proper to (e⁺e⁻) annihilation and DIS processes.

In fact, from fig. 3 we can work out the following quantities, which are

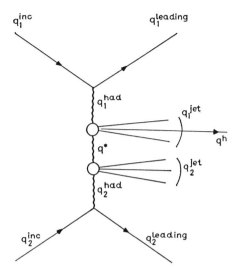

Fig. 3. – Schematic diagram for the (pp) interactions.

needed if we want to compare (pp) physics with (e⁺e⁻), *i.e.*

$$(11) \qquad (q_{\text{tot}}^{\text{had}})_{\text{pp}} = q_1^{\text{inc}} + q_2^{\text{inc}} - q_1^{\text{leading}} - q_2^{\text{leading}} = (q_1^{\text{had}} + q_2^{\text{had}})_{\text{pp}};$$

and (see formula (5))

$$(12) \qquad \sqrt{(q_{\text{tot}}^{\text{had}})_{\text{pp}}^2} = (\sqrt{s})_{\text{e}^+\text{e}^-}.$$

Moreover,

$$(13) \qquad (x)_{\text{pp}}^{\text{had}} = 2 \frac{q^{\text{h}} \cdot q_{\text{tot}}^{\text{had}}}{q_{\text{tot}}^{\text{had}} \cdot q_{\text{tot}}^{\text{had}}},$$

to be compared with

$$(14) \qquad (x)_{\text{e}^+\text{e}^-}^{\text{had}} = 2 \frac{q^{\text{h}} \cdot (q_{\text{tot}}^{\text{had}})_{\text{e}^+\text{e}^-}}{(q_{\text{tot}}^{\text{had}})_{\text{e}^+\text{e}^-} \cdot (q_{\text{tot}}^{\text{had}})_{\text{e}^+\text{e}^-}}.$$

The subscripts (e⁺e⁻) in formula (14) are there to make it clear that these quantities are measured in (e⁺e⁻) collisions and are the quantities equivalent to those measured in (pp) interactions.

The same (pp) diagram (fig. 3) can be used to work out the key quantities

6   M. BASILE, G. BONVICINI, G. CARA ROMEO, ETC.

needed when we want to compare (pp) physics with DIS. In this case we have

$$(15) \qquad (W^2)^{\text{had}}_{\text{pp}} = (q^{\text{had}}_1 + q^{\text{inc}}_2)^2$$

and

$$(16) \qquad (z)^{\text{had}}_{\text{pp}} = \frac{q^{\text{h}} \cdot q^{\text{inc}}_2}{q^{\text{had}}_1 \cdot q^{\text{inc}}_2}.$$

Note that in W^2 the leading proton No. 2 is not subtracted. This is the reason for the differences found in the comparison between DIS data and e$^+$e$^-$ (see sect. **3** and ref. ([15])). In fact W^2 is not the effective total energy available for particle production, owing to the presence there of the leading proton.

3. – Experimental results.

A series of experimental results, where (pp) interactions have been analysed *à la* e$^+$e$^-$ and *à la* DIS, have given impressive analogies in the multiparticle systems produced in these—so far considered—basically different processes: (pp), (e$^+$e$^-$), DIS.

The experimental data in which (pp) interactions are compared with (e$^+$e$^-$) are shown in fig. 4-18 and ref. ([1-14]).

([1]) M. BASILE, G. CARA ROMEO, L. CIFARELLI, A. CONTIN, G. D'ALÍ, P. DI CESARE, B. ESPOSITO, P. GIUSTI, T. MASSAM, F. PALMONARI, G. SARTORELLI, G. VALENTI and A. ZICHICHI: *Phys. Lett. B*, **92**, 367 (1980).

([2]) M. BASILE, G. CARA ROMEO, L. CIFARELLI, A. CONTIN, G. D'ALÍ, P. DI CESARE, B. ESPOSITO, P. GIUSTI, T. MASSAM, R. NANIA, F. PALMONARI, G. SARTORELLI, G. VALENTI and A. ZICHICHI: *Nuovo Cimento A*, **58**, 193 (1980).

([3]) M. BASILE, G. CARA ROMEO, L. CIFARELLI, A. CONTIN, G. D'ALÍ, P. DI CESARE, B. ESPOSITO, P. GIUSTI, T. MASSAM, R. NANIA, F. PALMONARI, G. SARTORELLI, G. VALENTI and A. ZICHICHI: *Phys. Lett. B*, **95**, 311 (1980).

([4]) M. BASILE, G. CARA ROMEO, L. CIFARELLI, A. CONTIN, G. D'ALÍ, P. DI CESARE, B. ESPOSITO, P. GIUSTI, T. MASSAM, R. NANIA, F. PALMONARI, G. SARTORELLI, G. VALENTI and A. ZICHICHI: *Lett. Nuovo Cimento*, **29**, 491 (1980).

([5]) M. BASILE, G. CARA ROMEO, L. CIFARELLI, A. CONTIN, G. D'ALÍ, P. DI CESARE, B. ESPOSITO, P. GIUSTI, T. MASSAM, R. NANIA, F. PALMONARI, G. SARTORELLI, M. SPINETTI, G. SUSINNO, G. VALENTI and A. ZICHICHI: *Phys. Lett. B*, **99**, 247 (1981).

([6]) M. BASILE, G. CARA ROMEO, L. CIFARELLI, A. CONTIN, G. D'ALÍ, P. DI CESARE, B. ESPOSITO, P. GIUSTI, T. MASSAM, R. NANIA, F. PALMONARI, G. SARTORELLI, M. SPINETTI, G. SUSINNO, G. VALENTI and A. ZICHICHI: *Lett. Nuovo Cimento*, **30**, 389 (1981).

([7]) M. BASILE, G. CARA ROMEO, L. CIFARELLI, A. CONTIN, G. D'ALÍ, P. DI CESARE,

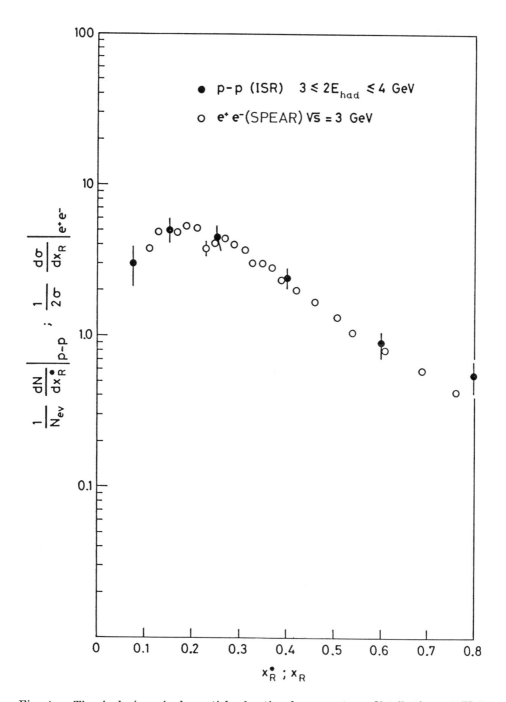

Fig. 4. – The inclusive single-particle fractional momentum distributions $(1/N_{\rm ev})\cdot$ $\cdot(dN/dx_{\rm R}^*)$ in the interval $3\,{\rm GeV}\leqslant 2E^{\rm had}\leqslant 4\,{\rm GeV}$ obtained from data at $\sqrt{s}=30\,{\rm GeV}$. Also shown are data from MARK I at SPEAR.

8   M. BASILE, G. BONVICINI, G. CARA ROMEO, ETC.

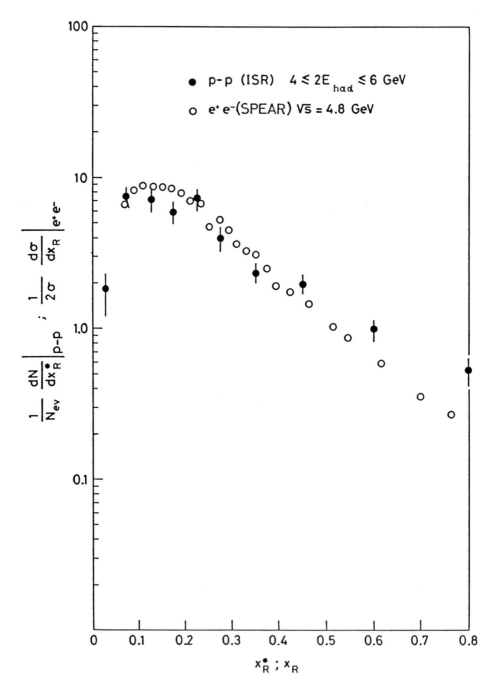

Fig. 5. – The inclusive single-particle fractional momentum distributions $(1/N_{\mathrm{ev}}) \cdot$ $\cdot (\mathrm{d}N/\mathrm{d}x_{\mathrm{R}}^*)$ in the interval $4\ \mathrm{GeV} < 2E^{\mathrm{had}} < 6\ \mathrm{GeV}$ obtained from data at $\sqrt{s} = 30\ \mathrm{GeV}$. Also shown are data from MARK I at SPEAR.

UNIVERSALITY FEATURES IN (pp), (e⁺e⁻) ETC.   **9**

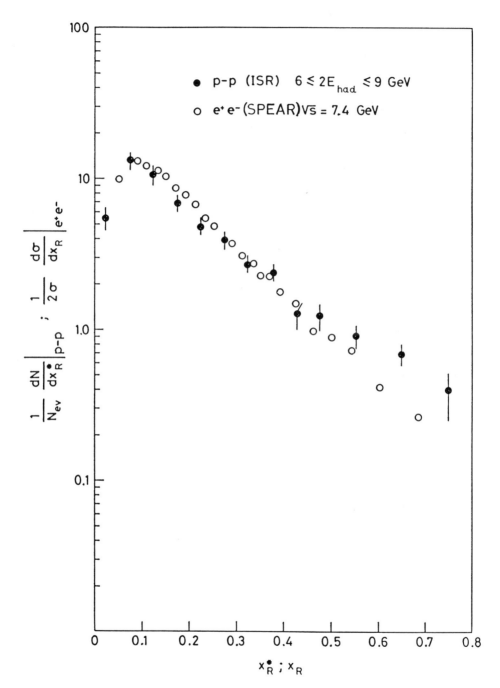

Fig. 6. – The inclusive single-particle fractional momentum distributions $(1/N_{\mathrm{ev}})\cdot$ $\cdot(\mathrm{d}N/\mathrm{d}x_{\mathbf{R}}^{*})$ in the interval $6\ \mathrm{GeV} \leqslant 2E^{\mathrm{had}} \leqslant 9\ \mathrm{GeV}$ obtained from data at $\sqrt{s} = 30\ \mathrm{GeV}$. Also shown are data from MARK I at SPEAR.

10   M. BASILE, G. BONVICINI, G. CARA ROMEO, ETC.

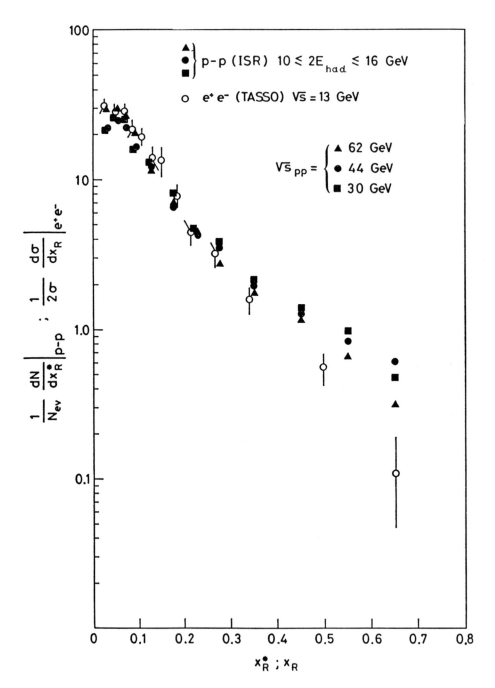

Fig. 7. – The inclusive single-particle fractional momentum distributions $(1/N_{ev}) \cdot$
$\cdot (dN/dx_R^*)$ in the interval $10 \text{ GeV} \leqslant 2E^{\text{had}} \leqslant 16 \text{ GeV}$ obtained from data at different
$(\sqrt{s})_{pp}$. Also shown are data from TASSO at PETRA.

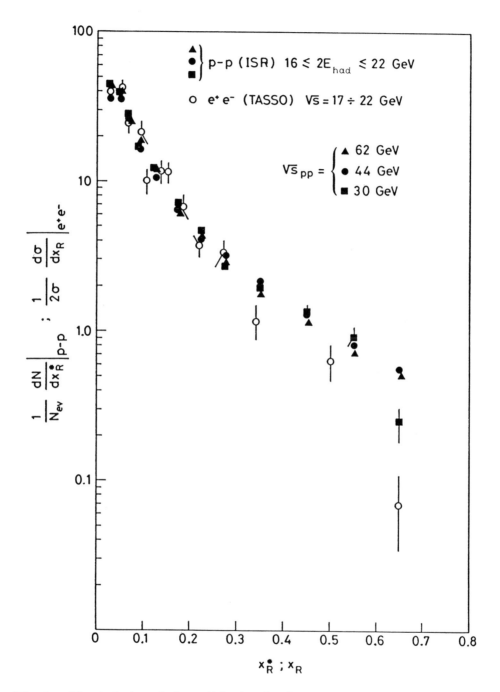

Fig. 8. – The inclusive single-particle fractional momentum distributions $(1/N_{\rm ev})\cdot$ $\cdot({\rm d}N/{\rm d}x_{\rm R}^*)$ in the interval $16\,{\rm GeV} \leqslant 2E^{\rm had} \leqslant 22\,{\rm GeV}$ obtained from data at different $(\sqrt{s})_{\rm pp}$. Also shown are data from TASSO at PETRA.

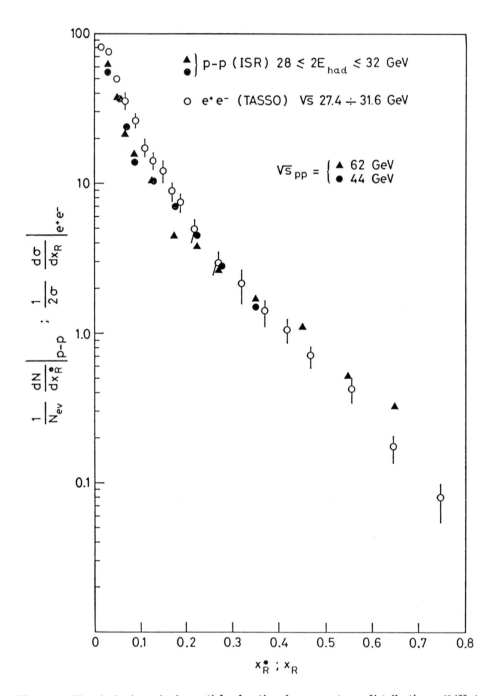

Fig. 9. – The inclusive single-particle fractional momentum distributions $(1/N_{ev})\cdot$ $\cdot(dN/dx_R^*)$ in the interval $28\,\text{GeV} \leqslant 2E^{\text{had}} \leqslant 32\,\text{GeV}$ obtained from data at different $(\sqrt{s})_{\text{pp}}$. Also shown are data from TASSO at PETRA.

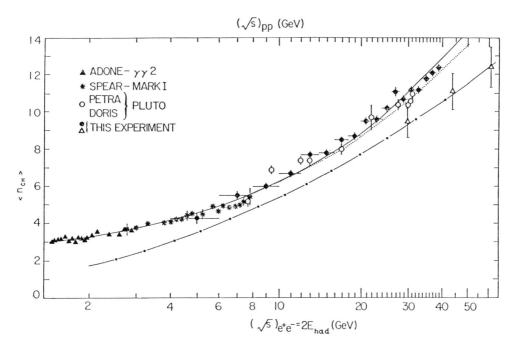

Fig. 10. – Mean charged-particle multiplicity (averaged over different $(\sqrt{s})_{pp}$) vs. $2E^{had}$, compared with (e⁺e⁻) data. The continuous line is the best fit to our data according to the formula $\langle n_{ch} \rangle = a + b \exp[c\sqrt{\ln (s/\Lambda)^2}]$. The dotted line is the best fit using PLUTO data. The dash-dotted line is the standard (pp) total charged-particle multiplicity with, superimposed, our data as open triangular points.

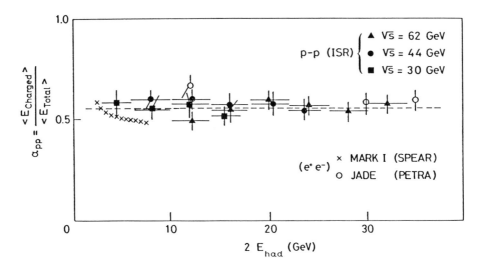

Fig. 11. – The charged-to-total energy ratio obtained in (pp) collisions α_{pp}, plotted vs. $2E^{had}$ and compared with (e⁺e⁻) obtained at SPEAR and PETRA.

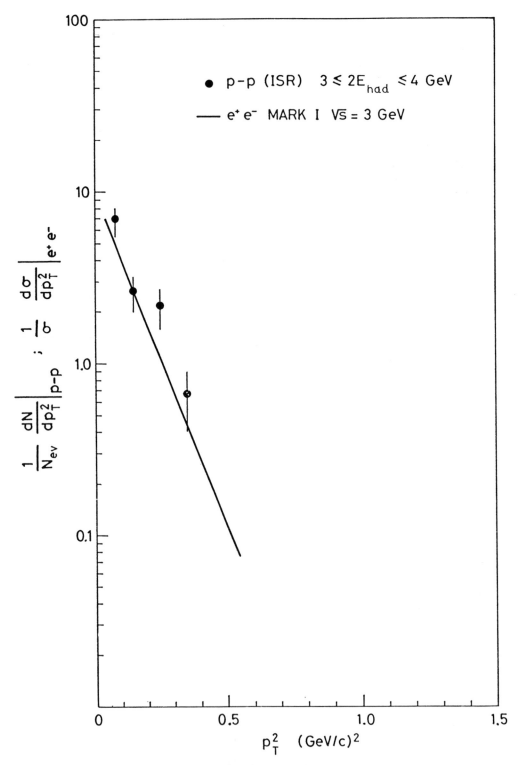

Fig. 12. – The inclusive single-particle transverse momentum distribution $(1/N_{ev}) \cdot$ $\cdot (dN/dp_T^2)$ in the interval $3\,\text{GeV} < 2E^{\text{had}} < 4\,\text{GeV}$ obtained from data at $\sqrt{s} = 30\,\text{GeV}$. Also shown is the fit to the SPEAR data (continuous line).

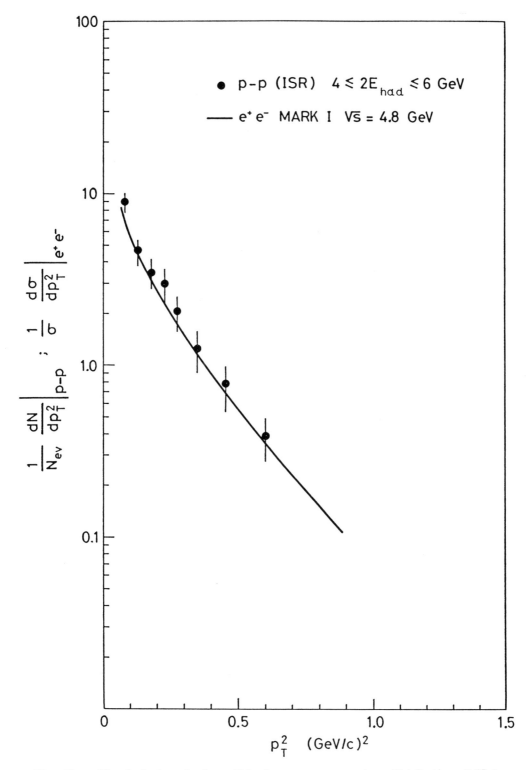

Fig. 13. – The inclusive single-particle transverse-momentum distribution $(1/N_{ev})\cdot$
$\cdot (\mathrm{d}N/\mathrm{d}p_T^2)$ in the interval $4\ \mathrm{GeV} \leqslant 2E^{\mathrm{had}} \leqslant 6\ \mathrm{GeV}$ obtained from data at $\sqrt{s} = 30\ \mathrm{GeV}$.
Also shown is the fit to the SPEAR data (continuous line).

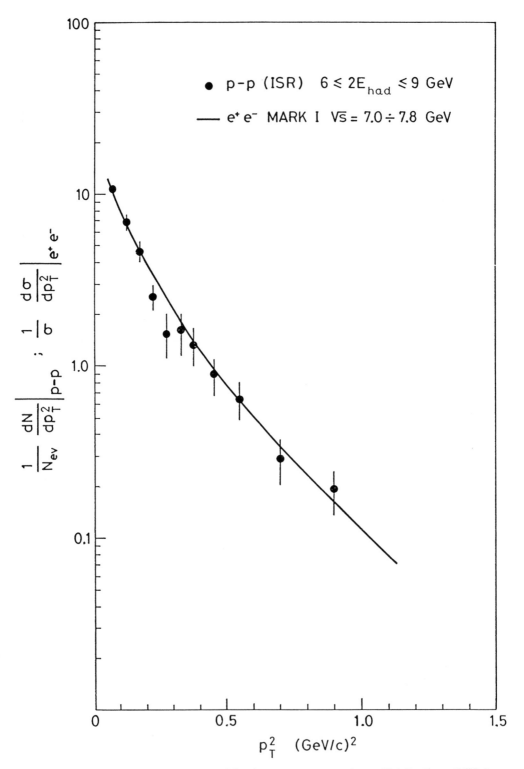

Fig. 14. – The inclusive single-particle transverse-momentum distribution $(1/N_{ev})\cdot$ $\cdot (dN/dp_T^2)$ in the interval $6\,\text{GeV} \leqslant 2E^{\text{had}} \leqslant 9\,\text{GeV}$ obtained from data at $\sqrt{s} = 30\,\text{GeV}$. Also shown is the fit to the SPEAR data (continuous line).

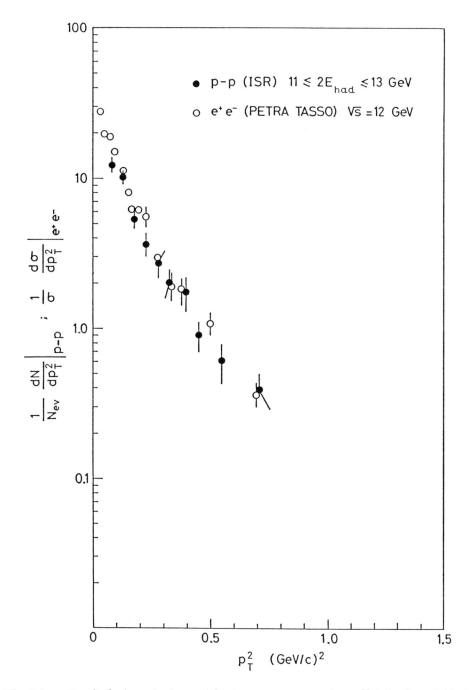

Fig. 15. – The inclusive single-particle transverse-momentum distributions $(1/N_{ev}) \cdot$
$\cdot (dN/dp_T^2)$ in the interval $11\,\mathrm{GeV} \leqslant 2E^{\mathrm{had}} \leqslant 13\,\mathrm{GeV}$ obtained from data at $\sqrt{s} = 62\,\mathrm{GeV}$.
Also shown are data from TASSO at PETRA.

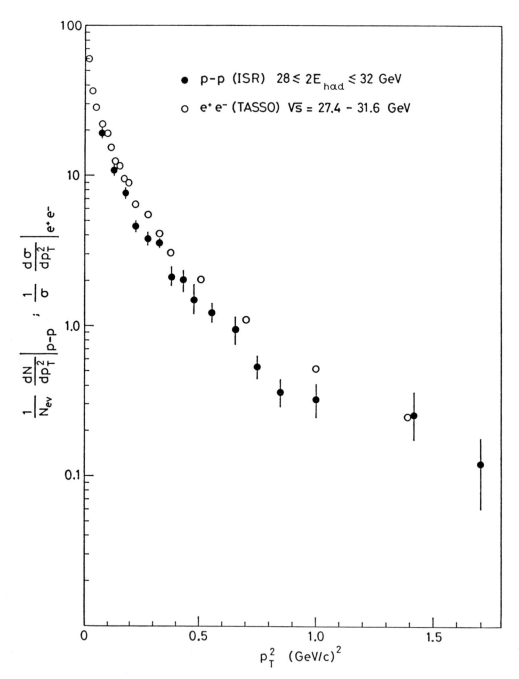

Fig. 16. – The inclusive single-particle transverse-momentum distributions $(1/N_{ev}) \cdot$ $\cdot (dN/dp_T^2)$ in the interval $28 \text{ GeV} \leqslant 2E^{had} \leqslant 32 \text{ GeV}$ obtained from data at $\sqrt{s} = 62 \text{ GeV}$ Also shown are data from TASSO at PETRA.

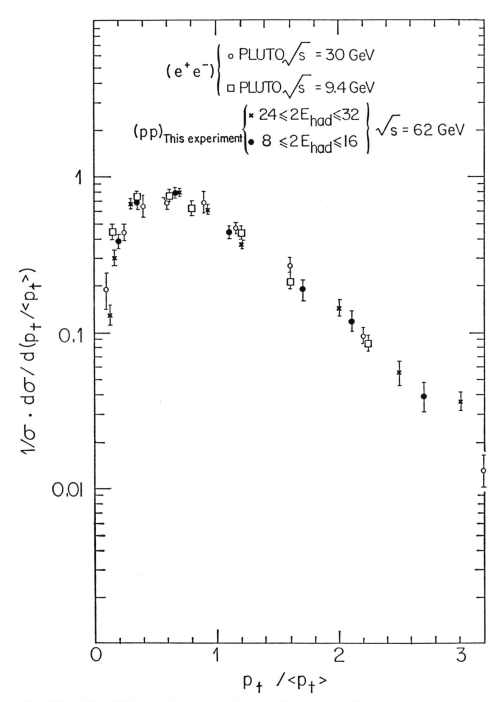

Fig. 17. – The differential cross-section $(1/\sigma)[\mathrm{d}\sigma/\mathrm{d}(p_t/\langle p_t\rangle)]$ *vs.* the «reduced» variable $p_t/\langle p_t\rangle$. These distributions allow a comparison of the multiparticle systems produced in (e⁺e⁻) annihilation and in (pp) interactions in terms of the renormalized transverse-momentum properties.

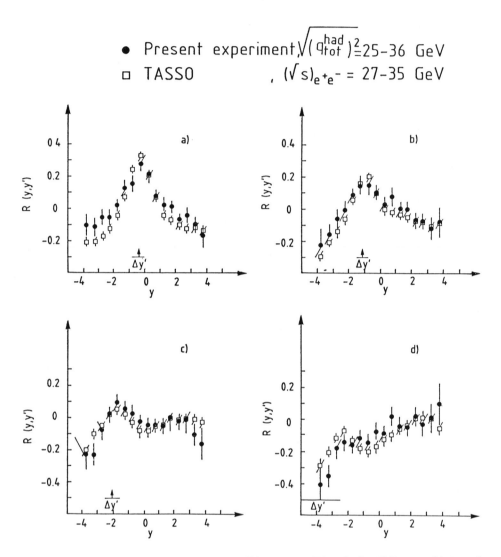

Fig. 18. – Two-particle correlation in rapidity space: $R(y, y')$, for different y' intervals, as measured in the present experiment after the leading-proton subtraction in the $\sqrt{(q_{tot}^{had})^2}$ range 25 to 36 GeV (black points), compared with the results by the TASSO collaboration at $(\sqrt{s})_{e^+e^-}$ between 27 and 35 GeV (open squares).

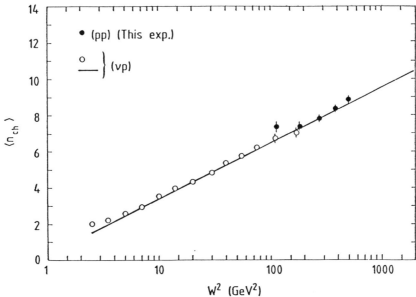

Fig. 19. – The average charged-particle multiplicities $\langle n_{ch} \rangle$ measured in (pp) at $(\sqrt{s})_{pp} = 30$ GeV, using a DIS-like analysis, are plotted *vs.* W^2 (black points). The open points are the (vp) data and the continuous line is their best fit.

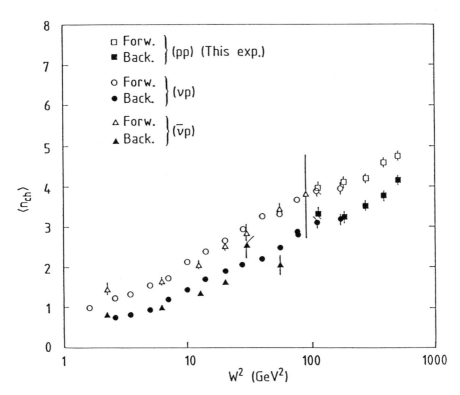

Fig. 20. – The mean charged-particle multiplicities $\langle n_{ch} \rangle_{F,B}$ in the forward and backward hemispheres *vs.* W^2, in (vp), ($\bar{v}$p) and (pp) interactions.

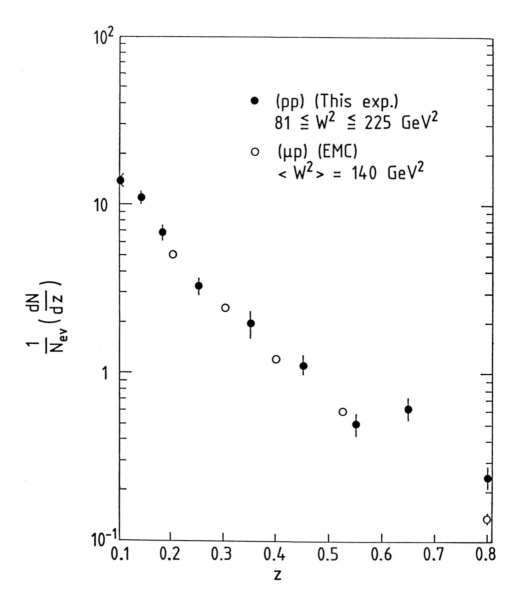

Fig. 21. – The inclusive distribution of the fractional energy z for (pp) reactions in the energy interval 81 (GeV)² $< W^2 <$ 225 (GeV)² compared with the data from (μp) reactions at $\langle W^2 \rangle$ = 140 (GeV)².

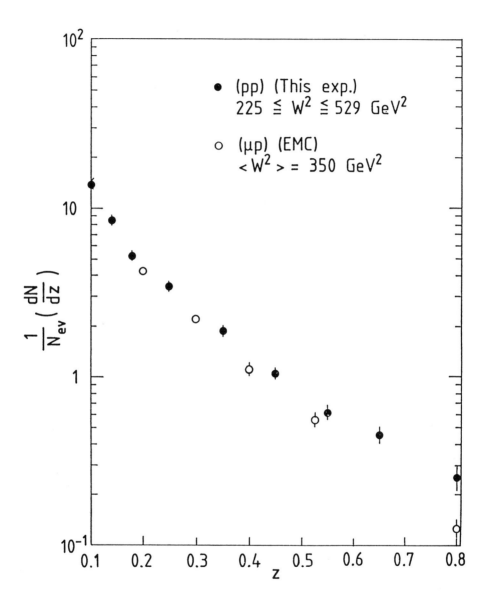

Fig. 22. – The inclusive distribution of the fractional energy z for (pp) reactions in the energy interval 225 (GeV)$^2 \leqslant W^2 \leqslant$ 529 (GeV)^2, compared with the data from (μp) reactions at $\langle W^2 \rangle = 350$ (GeV)^2.

The experimental data in which (pp) interactions are compared with DIS are shown in fig. 19-22 and ref. [15-17].

These comparisons show striking analogies with respect to the following quantities:

     i) the inclusive fractional energy distribution of the produced particles [1,2,9,11] (see fig. 4-9, 21, 22);

     ii) the average charged-particle multiplicities [3,8,12,15,16] (see fig. 10, 19, 20);

B. ESPOSITO, P. GIUSTI, T. MASSAM, R. NANIA, F. PALMONARI, G. SARTORELLI, M. SPINETTI, G. SUSINNO, G. VALENTI, L. VOTANO and A. ZICHICHI: *Lett. Nuovo Cimento*, **31**, 273 (1981).

[8] M. BASILE, G. CARA ROMEO, L. CIFARELLI, A. CONTIN, G. D'ALÍ, P. DI CESARE, B. ESPOSITO, P. GIUSTI, T. MASSAM, R. NANIA, F. PALMONARI, V. ROSSI, G. SARTORELLI, M. SPINETTI, G. SUSINNO, G. VALENTI, L. VOTANO and A. ZICHICHI: *Nuovo Cimento A*, **65**, 400 (1981).

[9] M. BASILE, G. CARA ROMEO, L. CIFARELLI, A. CONTIN, G. D'ALÍ, P. DI CESARE, B. ESPOSITO, P. GIUSTI, T. MASSAM, R. NANIA, F. PALMONARI, V. ROSSI, G. SARTORELLI, M. SPINETTI, G. SUSINNO, G. VALENTI, L. VOTANO and A. ZICHICHI: *Nuovo Cimento A*, **65**, 414 (1981).

[10] M. BASILE, G. CARA ROMEO, L. CIFARELLI, A. CONTIN, G. D'ALÍ, P. DI CESARE, B. ESPOSITO, P. GIUSTI, T. MASSAM, R. NANIA, F. PALMONARI, V. ROSSI, G. SARTORELLI, M. SPINETTI, G. SUSINNO, G. VALENTI, L. VOTANO and A. ZICHICHI: *Lett. Nuovo Cimento*, **32**, 210 (1981).

[11] M. BASILE, G. CARA ROMEO, L. CIFARELLI, A. CONTIN, G. D'ALÍ, P. DI CESARE, B. ESPOSITO, P. GIUSTI, T. MASSAM, R. NANIA, F. PALMONARI, V. ROSSI, F. ROHRBACH, G. SARTORELLI, M. SPINETTI, G. SUSINNO, G. VALENTI, L. VOTANO and A. ZICHICHI: *Nuovo Cimento A*, **67**, 53 (1982).

[12] M. BASILE, G. BONVICINI, G. CARA ROMEO, L. CIFARELLI, A. CONTIN, M. CURATOLO, G. D'ALÍ, P. DI CESARE, B. ESPOSITO, P. GIUSTI, T. MASSAM, R. NANIA, F. PALMONARI, A. PETROSINO, V. ROSSI, G. SARTORELLI, M. SPINETTI, G. SUSINNO, G. VALENTI, L. VOTANO and A. ZICHICHI: *Nuovo Cimento A*, **67**, 244 (1982).

[13] M. BASILE, G. BONVICINI, G. CARA ROMEO, L. CIFARELLI, A. CONTIN, M. CURATOLO, G. D'ALÍ, C. DEL PAPA, B. ESPOSITO, P. GIUSTI, F. MASSAM, R. NANIA, F. PALMONARI, G. SARTORELLI, M. SPINETTI, G. SUSINNO, L. VOTANO and A. ZICHICHI: *Nuovo Cimento*, **73**, 329 (1983).

[14] G. BONVICINI, G. CARA ROMEO, L. CIFARELLI, A. CONTIN, M. CURATOLO, G. D'ALÍ, C. DEL PAPA, B. ESPOSITO, P. GIUSTI, T. MASSAM, R. NANIA, G. SARTORELLI, G. SUSINNO, L. VOTANO and A. ZICHICHI: *Lett. Nuovo Cimento*, **36**, 563 (1983).

[15] M. BASILE, G. BONVICINI, G. CARA ROMEO, L. CIFARELLI, A. CONTIN, M. CURATOLO, G. D'ALÍ, C. DEL PAPA, B. ESPOSITO, P. GIUSTI, T. MASSAM, R. NANIA, F. PALMONARI, G. SARTORELLI, G. SUSINNO, L. VOTANO and A. ZICHICHI: *Lett. Nuovo Cimento*, **36**, 303 (1983).

[16] G. BONVICINI, G. CARA ROMEO, L. CIFARELLI, A. CONTIN, M. CURATOLO, G. D'ALÍ, C. DEL PAPA, B. ESPOSITO, P. GIUSTI, T. MASSAM, R. NANIA, F. PALMONARI, G. SARTORELLI, G. SUSINNO, L. VOTANO and A. ZICHICHI: *Lett. Nuovo Cimento*, **36**, 555 (1983).

[17] G. BONVICINI, G. CARA ROMEO, L. CIFARELLI, A. CONTIN, M. CURATOLO, G. D'ALÍ, C. DEL PAPA, B. ESPOSITO, P. GIUSTI, T. MASSAM, R. NANIA, G. SARTORELLI, G. SUSINNO, L. VOTANO and A. ZICHICHI: *Lett. Nuovo Cimento*, **37**, 289 (1983).

iii) the ratio of the average energy associated with the charged particles over the total energy available for particle production ([5]) (see fig. 11);

iv) the inclusive transverse-momentum distribution of the produced particles ([7,10]) (see fig. 12-17);

v) the correlation functions in rapidity ([14]) (see fig. 18).

Notice the power of the (pp) interaction. Once this is analysed in the correct way, it produces results equivalent to (e⁺e⁻) and DIS.

This means that there is an important universality in these ways of producing multihadronic systems.

4. – Extrapolations to collider physics.

From the above analysis we can conclude that

i) the leading effects must be subtracted and the correct variables have to be used if we want to compare purely hadronic interactions with (e⁺e⁻) and DIS;

ii) the old myth, based on the belief that in order to compare (pp) with (e⁺e⁻) and DIS you need high-$p_{\rm T}$ (pp) interactions, is over. In fact we have proved that low-$p_{\rm T}$ (pp) interactions produce results in excellent agreement with (e⁺e⁻) annihilation and DIS processes, the basic parameter in (pp) interactions being $\sqrt{(q_{\rm tot}^{\rm had})^2}$ for a comparison with (e⁺e⁻), and $\sqrt{W_{\rm (pp)}^2}$ for a comparison with DIS.

The existence of high-$p_{\rm T}$ events means that pointlike constituents exist inside the nucleon. But low-$p_{\rm T}$ events contain the same amount of basic information as high-$p_{\rm T}$ events. The only difference is expected in $\langle p_t \rangle$. In fact our analysis of the inclusive transverse-momentum distribution, in terms of the renormalized variable $p_t/\langle p_t \rangle$ (notice that p_t indicates the transverse momentum of the particles produced with respect to the jet axis, and $p_{\rm T}$ with respect to the colliding (pp) or (p$\bar{\rm p}$) axis), is suggestive of a very interesting possibility: multiparticle systems produced at high $p_{\rm T}$ could show, at equivalent $\sqrt{(q_{\rm tot}^{\rm had})^2}$, higher values of $\langle p_t \rangle$. This should be the only difference between multiparticle systems with the same $\sqrt{(q_{\rm tot}^{\rm had})^2}$ produced at low $p_{\rm T}$ and high $p_{\rm T}$.

There are two ways of producing $\sqrt{(q_{\rm tot}^{\rm had})^2}$:

i) one is at low $p_{\rm T}$, and we have seen what happens;

ii) the other is at high $p_{\rm T}$: we have not been able to compare, at constant values of $\sqrt{(q_{\rm tot}^{\rm had})^2}$, the multiparticle systems produced in (pp) interactions at high $p_{\rm T}$ and low $p_{\rm T}$, the reason being the lack of CERN ISR time. However, as mentioned above, the agreement between our data and (e⁺e⁻) data in the variable $(p_t/\langle p_t \rangle)$, makes it possible to foresee what should change between low-p_T and high-p_T multiparticle jets.

Now we come to the extrapolation: *the extrapolation of our method to the CERN p$\bar{p}$ collider* [18,19] would allow a large energy jump and could produce clear evidence for or against our prediction. For example, if two jets at the p$\bar{p}$ collider are produced back-to-back with the same transverse energy E_T, then we have

$$\sqrt{(q_{tot}^{had})^2} \simeq 2E_T \, .$$

Suppose that we are at

$$2E_T = 100 \text{ GeV} \, .$$

This system, according to our extrapolation, should be like a multiparticle state produced by $(\sqrt{s})_{e^+e^-} = 100$ GeV.

However, there is a very important check to make using collider data, without the need for the (e⁺e⁻) data.

The key point is to see if, at the CERN p$\bar{p}$ collider, a multiparticle system produced at low p_T but with

$$\sqrt{(q_{tot}^{had})^2} = 100 \text{ GeV}$$

looks like the one produced at high E_T. The main difference we can expect is the value of $\langle p_t \rangle$.

To check these points is another important contribution to understanding hadron production at extreme energies.

5. – Conclusions.

The new method of studying (pp) and (p$\bar{p}$) collisions—based on the subtraction of the «leading» effects and the use of correct variables—allows us to put on equal footing the multiparticle systems that are produced in purely hadronic interactions, in (e⁺e⁻) annihilation and in DIS processes.

The implications of this method for the physics to be studied in future machines are of great interest.

Purely hadronic interactions means using machines such as the CERN Intersecting Storage Ring (ISR), the CERN p$\bar{p}$ collider, the BNL-CBA collider, and the FNAL p$\bar{p}$ collider.

[18] UA2 COLLABORATION: *Phys. Lett. B*, **118**, 203 (1982).
[19] UA1 COLLABORATION: preprint CERN-EP/83-02 (1983).

(e⁺e⁻) *annihilation* means using machines such as LEP and its possible developments.

DIS *processes* means using machines such as HERA.

The « leading » subtraction and the use of the correct variables allow us to show that universality features are present, in the production of multibody final states in (pp), (e⁺e⁻) and DIS.

● RIASSUNTO

L'uso delle variabili corrette, nei processi (pp), (e⁺e⁻) e scattering inelastico profondo (DIS), permette di mettere in evidenza l'esistenza di caratteristiche universali in tali interazioni, considerate sinora modi molto differenti di produrre stati multiadronici.

Универсальные особенности в (pp), (e⁺e⁻) и процессах глубоко неупругого рассеяния.

Резюме (*). — Использование правильных переменных в (p, p), (e⁺e⁻) и процессах глубоко неупругого рассеяния позволяет установить универсальные особенности в этих реакциях.

(*) *Переведено редакцией.*

SERVICE D'INFORMATION
SCIENTIFIQUE
CERN

M. BASILE, *et al.*
1 Gennaio 1984
Il Nuovo Cimento
Serie 11, Vol. 79 A, pag. 1-27

IL NUOVO CIMENTO   VOL. 73 A, N. 4   21 Febbraio 1983

Experimental Proof that the Leading Protons are not Correlated.

M. Basile, G. Bonvicini, G. Cara Romeo, L. Cifarelli, A. Contin
M. Curatolo, G. D'Alí, C. Del Papa, B. Esposito, P. Giusti, T. Massam
R. Nania, F. Palmonari, G. Sartorelli, G. Susinno
L. Votano and A. Zichichi

CERN - Geneva, Switzerland
Istituto di Fisica dell'Università - Bologna, Italia
Istituto Nazionale di Fisica Nucleare, Laboratori Nazionali di Frascati - Frascati, Italia
Istituto Nazionale di Fisica Nucleare - Sezione di Bologna, Italia
Istituto di Fisica dell'Università - Perugia, Italia

(ricevuto il 19 Novembre 1982)

Summary. — The correlations between the two « leading » protons in the x-range $0.4 \leqslant x < 0.9$ are measured to be below $\pm 1\%$, at the highest ISR energy.

PACS. 13.85. – Hadron-induced high- and super–high-energy interactions, energy > 10 GeV.

In previous papers we have reported [1-15] on the relevance of studying proton-proton interactions with the technique of subtracting the « leading »-proton effects.

[1] M. Basile, G. Cara Romeo, L. Cifarelli, A. Contin, G. D'Alí, P. Di Cesare, B. Esposito, P. Giusti, T. Massam, F. Palmonari, G. Sartorelli, G. Valenti and A. Zichichi: Phys. Lett. B, 92, 367 (1980).
[2] M. Basile, G. Cara Romeo, L. Cifarelli, A. Contin, G. D'Alí, P. Di Cesare, B. Esposito, P. Giusti, T. Massam, R. Nania, F. Palmonari, G. Sartorelli, G. Valenti and A. Zichichi: Nuovo Cimento A, 58, 193 (1980).
[3] M. Basile, G. Cara Romeo, L. Cifarelli, A. Contin, G. D'Alí, P. Di Cesare, B. Esposito, P. Giusti, T. Massam, R. Nania, F. Palmonari, G. Sartorelli, G. Va-

Here we report on the study of the correlations between the two leading protons, a study which is relevant in that it is the first step that has to be taken before trying to understand if there are similarities among the correlations

LENTI and A. ZICHICHI: *Phys. Lett. B*, **95**, 311 (1980).

(⁴) M. BASILE, G. CARA ROMEO, L. CIFARELLI, A. CONTIN, G. D'ALÍ, P. DI CESARE, B. ESPOSITO, P. GIUSTI, T. MASSAM, R. NANIA, F. PALMONARI, G. SARTORELLI, G. VALENTI and A. ZICHICHI: *Lett. Nuovo Cimento*, **29**, 491 (1980).

(⁵) M. BASILE, G. CARA ROMEO, L. CIFARELLI, A. CONTIN, G. D'ALÍ, P. DI CESARE, B. ESPOSITO, P. GIUSTI, T. MASSAM, R. NANIA, F. PALMONARI, G. SARTORELLI, M. SPINETTI, S. SUSINNO, G. VALENTI and A. ZICHICHI: *Phys. Lett. B*, **99**, 247 (1981).

(⁶) M. BASILE, G. CARA ROMEO, L. CIFARELLI, A. CONTIN, G. D'ALÍ, P. DI CESARE, B. ESPOSITO, P. GIUSTI, T. MASSAM, R. NANIA, F. PALMONARI, G. SARTORELLI, M. SPINETTI, G. SUSINNO, G. VALENTI and A. ZICHICHI: *Lett. Nuovo Cimento*, **30**, 389 (1981).

(⁷) M. BASILE, G. CARA ROMEO, L. CIFARELLI, A. CONTIN, G. D'ALÍ, P. DI CESARE, B. ESPOSITO, P. GIUSTI, T. MASSAM, R. NANIA, F. PALMONARI, G. SARTORELLI, M. SPINETTI, G. SUSINNO, G. VALENTI, L. VOTANO and A. ZICHICHI: *Lett. Nuovo Cimento*, **31**, 273 (1981).

(⁸) M. BASILE, G. CARA ROMEO, L. CIFARELLI, A. CONTIN, G. D'ALÍ, P. DI CESARE, B. ESPOSITO, P. GIUSTI, T. MASSAM, R. NANIA, F. PALMONARI, V. ROSSI, G. SARTORELLI, M. SPINETTI, G. SUSINNO, G. VALENTI, L. VOTANO and A. ZICHICHI: *Nuovo Cimento A*, **65**, 400 (1981).

(⁹) M. BASILE, G. CARA ROMEO, L. CIFARELLI, A. CONTIN, G. D'ALÍ, P. DI CESARE, B. ESPOSITO, P. GIUSTI, T. MASSAM, R. NANIA, F. PALMONARI, V. ROSSI, G. SARTORELLI, M. SPINETTI, G. SUSINNO, G. VALENTI, L. VOTANO and A. ZICHICHI: *Nuovo Cimento A*, **65**, 414 (1981).

(¹⁰) M. BASILE, G. CARA ROMEO, L. CIFARELLI, A. CONTIN, G. D'ALÍ, P. DI CESARE, B. ESPOSITO, P. GIUSTI, T. MASSAM, R. NANIA, F. PALMONARI, V. ROSSI, G. SARTORELLI, M. SPINETTI, G. SUSINNO, G. VALENTI, L. VOTANO and A. ZICHICHI: *Lett. Nuovo Cimento*, **32**, 210 (1981).

(¹¹) M. BASILE, G. CARA ROMEO, L. CIFARELLI, A. CONTIN, G. D'ALÍ, P. DI CESARE, B. ESPOSITO, P. GIUSTI, T. MASSAM, R. NANIA, F. PALMONARI, V. ROSSI, G. SARTORELLI, M. SPINETTI, G. SUSINNO, G. VALENTI, L. VOTANO and A. ZICHICHI: *Nuovo Cimento A*, **66**, 129 (1981).

(¹²) M. BASILE, G. CARA ROMEO, L. CIFARELLI, A. CONTIN, G. D'ALÍ, P. DI CESARE, B. ESPOSITO, P. GIUSTI, T. MASSAM, R. NANIA, F. PALMONARI, V. ROSSI, G. SARTORELLI, M. SPINETTI, G. SUSINNO, G. VALENTI, L. VOTANO and A. ZICHICHI: *Lett. Nuovo Cimento*, **32**, 321 (1981).

(¹³) M. BASILE, G. CARA ROMEO, L. CIFARELLI, A. CONTIN, G. D'ALÍ, P. DI CESARE, B. ESPOSITO, P. GIUSTI, T. MASSAM, R. NANIA, F. PALMONARI, V. ROSSI, F. ROHRBACH, G. SARTORELLI, M. SPINETTI, G. SUSINNO, G. VALENTI, L. VOTANO and A. ZICHICHI: *Nuovo Cimento A*, **67**, 53 (1982).

(¹⁴) M. BASILE, G. BONVICINI, G. CARA ROMEO, L. CIFARELLI, A. CONTIN, M. CURATOLO, G. D'ALÍ, P. DI CESARE, B. ESPOSITO, P. GIUSTI, T. MASSAM, R. NANIA, F. PALMONARI, A. PETROSINO, V. ROSSI, G. SARTORELLI, M. SPINETTI, G. SUSINNO, G. VALENTI, L. VOTANO and A. ZICHICHI: *Nuovo Cimento A*, **67**, 244 (1982).

(¹⁵) M. BASILE, G. CARA ROMEO, L. CIFARELLI, A. CONTIN, M. CURATOLO, G. D'ALÍ, P. DI CESARE, B. ESPOSITO, P. GIUSTI, T. MASSAM, R. NANIA, F. PALMONARI, A. PETROSINO, V. ROSSI, G. SARTORELLI, M. SPINETTI, G. SUSINNO, G. VALENTI, L. VOTANO and A. ZICHICHI: preprint CERN-EP/81-147 (1981).

existing in multiparticle hadronic systems produced in (e⁺e⁻) annihilations and in (pp) interactions [16].

The correlations studied in the (pp) channels have so far [16] been made without the subtraction of the leading protons. Nevertheless, conclusions have been proposed [16] in terms of differences existing between (pp) and (e⁺e⁻) cases.

We have reported in previous papers [1-15] that the « leading »-particle effect is a basic feature of hadronic physics. We, therefore, believe that the first step in trying to understand the comparison between (pp) and (e⁺e⁻) is to start with the key phenomenon that governs the multiparticle hadronic systems produced in (pp) interactions, *i.e.* the study of the correlation between the two leading protons. In fact the basic quantity which, according to our studies [1-15], governs the multiparticle production process in (pp) interaction is the total effective hadronic energy available. This quantity is obtained by subtracting, from the total invariant four-momentum of the initial state, the total invariant four-momentum carried out by the two leading protons.

Let us recall the main points of this argument. Given two protons in the initial state, let q_1^{inc} and q_2^{inc} be their four-momenta. The total invariant four-momentum of the two colliding protons is

$$(1) \qquad q_{\text{total}}^{\text{inc}} = q_1^{\text{inc}} + q_2^{\text{inc}} ,$$

and the invariant mass of the system is

$$(2) \qquad \sqrt{(q_{\text{total}}^{\text{inc}})^2} = 2 E_{1,2}^{\text{inc}} ,$$

where $E_1^{\text{inc}} = E_2^{\text{inc}} = E^{\text{beam}}$ in the (pp) c.m. system. In the standard notation

$$(3) \qquad \sqrt{(q_{\text{total}}^{\text{inc}})^2} = (\sqrt{s})_{\text{pp}} .$$

Let q_1^{leading} and q_2^{leading} be the invariant four-momenta of the two leading protons in the final state. The total invariant four-momentum carried away by the two leading protons will be

$$(4) \qquad q_1^{\text{leading}} + q_2^{\text{leading}} = q_{\text{total}}^{\text{leading}} .$$

This is the basic quantity to subtract from $q_{\text{total}}^{\text{inc}}$ in order to know the effective total hadronic energy available for particle production in a (pp) collision:

$$(5) \qquad \sqrt{(q_{\text{total}}^{\text{had}})^2} = \sqrt{(q_{\text{total}}^{\text{inc}} - q_{\text{total}}^{\text{leading}})^2} .$$

---

[16] For a review on and references to the original work, see W. KOCH: *Proceedings of the XIII International Symposium on Multiparticle Dynamics* (Volendam, 1982), to be published.

332   M. BASILE, G. BONVICINI, G. CARA ROMEO, ETC.

The main point of the present analysis is to study if the two leading protons
are correlated.

The data used in the present analysis have been taken at the CERN Intersecting Storage Rings (ISR) using a system of multiwire proportional chambers
in a large-volume magnetic field ([1-15]). The reaction studied was

$$(6) \qquad p_1^{inc} + p_2^{inc} \rightarrow p_1^{leading} + p_2^{leading} + anything,$$

where $p_{1,2}^{inc}$ indicate the two incident protons, and $p_{1,2}^{leading}$ the two leading protons.
The data taking was performed by using unbiased events in order to have a
set of genuine inclusive (pp) interactions (reaction (6)).

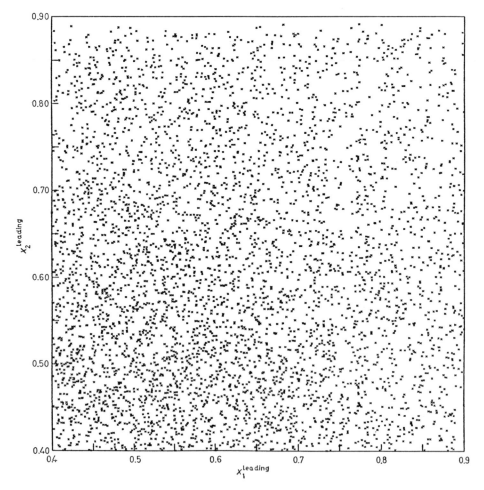

Fig. 1. – Scatter plot of the fractional energies of the two leading protons,
$x_{1,2}^{leading} = E_{1,2}^{leading}/E_{1,2}^{inc}$, in the range $0.4 \leqslant x_{1,2}^{leading} < 0.9$, at the (pp) c.m. energy
$(\sqrt{s})_{pp} = 62$ GeV.

The results are presented in fig. 1, as a scatter plot in terms of the quantities $x_{1,2}^{\text{leading}} = E_{1,2}^{\text{leading}}/E_{1,2}^{\text{inc}}$. Each point in the $(x_1^{\text{leading}}, x_2^{\text{leading}})$ scatter plot represents an event. The uniformity of the distribution of the events in the scatter plot is the proof that there is no correlation between the two leading

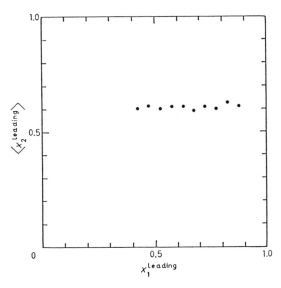

Fig. 2. – The average value $\langle x_2^{\text{leading}}\rangle$ for fixed x_1^{leading} is plotted *vs.* x_1^{leading}. This is obtained from the scatter plot of fig. 1.

protons. A more quantitative analysis of the same data is reported in fig. 2, where the average value $\langle x_2^{\text{leading}}\rangle$ at fixed x_1

$$[\langle x_2^{\text{leading}}\rangle]_{x_1 \text{ fixed}} = \left[\frac{\sum\limits_{i=1}^{N} x_2^{\text{leading}}}{N}\right]_{x_1 \text{ fixed}}$$

is plotted *vs.* x_1. The data correspond to a total of 4841 (pp) events at $(\sqrt{s})_{\text{pp}} = 62$ GeV.

The basic features of the data presented in fig. 2 are:

   i) All $\langle x_2^{\text{leading}}\rangle$ values are known with a precision which ranges from 1% to 1.5%. Within their statistical error, all $\langle x_2^{\text{leading}}\rangle$ values are compatible with a constant value, their average.

   ii) The maximum slope, compatible with the uncertainty of all data, corresponds to a maximum variation in $\langle x_2^{\text{leading}}\rangle$ of $\pm 1\%$.

Any correlation between the two leading protons, should it be there, must be so small that it does not produce effects above the quoted $\pm 1\%$ level.

The fact that the two leading protons were not correlated is a result that we had already obtained during the very early stage of our studies. It has been presented at many seminars and conferences but never published, because, in our new way of analysing multiparticle hadronic systems produced in (pp) collisions, the absence of a correlation between the two leading protons was the starting point. If strong correlations had been present between the leading protons, our analysis would have lost its very simple nature and would have needed involved and complicated arguments.

The results of our studies on two-body correlations [17] and on charge correlations [18] among the particles produced in (pp) interactions, once the « leading protons » have been subtracted, will be reported elsewhere [17,18].

* * *

We thank Prof. G. WOLF for having suggested to us the publication of the leading-proton correlation studies. We were not aware of the fact that this basic feature of our new way of analysing multiparticle hadronic systems produced in (pp) interactions was not so well known.

[17] M. BASILE, G. BONVICINI, G. CARA ROMEO, L. CIFARELLI, A. CONTIN, M. CURATOLO, G. D'ALÍ, C. DEL PAPA, B. ESPOSITO, P. GIUSTI, T. MASSAM, R. NANIA, F. PALMONARI, G. SARTORELLI, M. SPINETTI, G. SUSINNO, L. VOTANO and A. ZICHICHI: *Comparison of two-particle correlations in* (pp) *interactions and* (e+e−) *annihilation*, in preparation.
[18] M. BASILE, G. BONVICINI, G. CARA ROMEO, L. CIFARELLI, A. CONTIN, M. CURATOLO, G. D'ALÍ, C. DEL PAPA, B. ESPOSITO, P. GIUSTI, T. MASSAM, R. NANIA, F. PALMONARI, G. SARTORELLI, M. SPINETTI, G. SUSINNO, L. VOTANO and A. ZICHICHI: *Charge correlation studies in* (pp) *compared with* (e+e−), in preparation.

● RIASSUNTO

Nelle interazioni protone-protone alla piú alta energia degli ISR, la correlazione in x tra i due protoni «leading» misurata nell'intervallo $0.4 \leqslant x \leqslant 0.9$ è stata trovata inferiore a $\pm 1\%$.

Резюме не получено.

CERN
SERVICE D'INFORMATION
SCIENTIFIQUE

M. BASILE, *et al.*
21 Febbraio 1983
Il Nuovo Cimento
Serie 11, Vol. 73 A, pag. 329-334

OLD AND NEW FORCES OF NATURE
Edited by Antonino Zichichi
(Plenum Publishing Corporation, 1989)

UNIVERSALITY PROPERTIES IN NON–PERTURBATIVE QCD

A. Zichichi

CERN, Geneva, Switzerland

1. INTRODUCTION: QCD AND THE LEADING EFFECT

If QCD exists, it should show universality properties, both in the perturbative and non–perturbative channels. The content of this lecture will be to show that QCD has universality features in the non–perturbative channels.

Since ever, strong interactions have had a difficult time to show universality features. Before QCD started, the story concerning strong interactions was completely different from what we see now. At that time we had $SU(3)_{flavour}$, the famous global symmetry of Gell–Mann and Ne'eman. Now we work with a symmetry $SU(3)_{colour}$, which is a local one. And it is an exact symmetry: the gluons are gauge bosons and remain massless to all orders of perturbation theory.

Let me remind you that QCD has progressed from a candidate theory of the strong interactions to an essential ingredient of the so–called Standard Model: $SU(3) \times SU(2) \times U(1)$, characterized by three coupling constants: α_s, α_W, and α_{EM}. Notice that α_W and α_{EM} can be determined from low energy experiments, due to the use of weak coupling perturbation theory. The asymptotic freedom of QCD, however, yields a coupling constant α_s that is small only at very high energies. So only a selected type of experiments can be useful for the determination of α_s. These are the experiments sensitive to the dynamics at short distances.

One important parameter in this dynamics, is the scale which determines the rate at which α_s runs, and this is the meaning of the quantity Λ. This quantity remains important even when the Standard Model is incorporated into a GUT. For example, using $SU(5)$, the values of Λ become connected to the weak mixing angle and to the masses of W and Z.

The following table shows an example (Ref. 1)

Λ(MeV)	$\sin^2\theta_W$	$M_{W^\pm}$(GeV/c^2)	M_{Z^0}(GeV/c^2)
25	0.224	81.3	92.3
50	0.220	82.0	92.8
100	0.216	82.8	93.6
200	0.212	83.5	94.3
400	0.208	84.3	94.9

Figure 1 illustrates this connection, i.e. the values of the W and Z masses obtained using different values of Λ. The results obtained at CERN (Ref. 2) are shown for comparison. Notice the remarkable agreement.

For QCD to be a serious candidate theory able to describe the quark gluon interactions, it must be possible to describe all the reactions in terms of a single dimensional parameter Λ. The present success of QCD in the perturbative level lies in a universal determination of Λ from apparently very different reactions, like, for example, deep inelastic scattering; leptonic, hadronic and photonic widths of heavy quarkonia, such as J/ψ and Υ; jets and e^+e^- annihilation; and structure functions of the photon. However, all these tests are done at the perturbative level. This means that one has to find kinematic domains where perturbation theory applies and to insist on some inclusive measurements that do not involve details of single hadron distributions or correlations.

It is not always possible to fulfill these requirements of perturbative QCD. Nevertheless, whenever these requirements are fulfilled, one gets values of Λ which typically lie in the 100–200 MeV range (Ref. 1). Figure 2 shows a summary of the Λ values. These include results from fragmentation in e^+e^-, the branching ratios of hadrons into leptons for the Υ, the J/ψ and η_c hyperfine splitting, deep inelastic scattering, $q\bar{q}$ potentials in QCD, QCD sum rules, and photon structure functions. In spite of the fact that the values of Λ fall in a not too large interval, QCD is far from being a well established formulation of the interactions between quarks and gluons. There are a lot of things to understand and to prove experimentally.

The ultimate test of QCD involves universality at the non–perturbative level. This means the universality of the coupling constant α_s and the universal space–time evolution of hadronic states after high energy collisions. If QCD is to be a theory of the basic strong interactions i.e., a theory of the interactions between quarks and gluons, it has to show universality features. What is meant by the universal evolution (space–time) of hadronic states is that the fragmentation functions must be independent of their environment. The study of time–dependent hadronization phenomena is not yet amenable, using known and tested theoretical tools. It is here that the experimental analysis can play a leading role.

Before our new method (Ref. 3) of studying multiparticle hadronic systems was invented, QCD processes of non–perturbative nature carried no sign of universality. In other words, the experimental data seemed to show that the space–time evolution of hadronic states was linked with the "environment". And this means that weak, electromagnetic and strong processes seemed to produce different results for hadronic multiplicities, for transverse momentum distributions, for inclusive hadronic distributions, for two particles correlations etc. Moreover, strong interactions processes

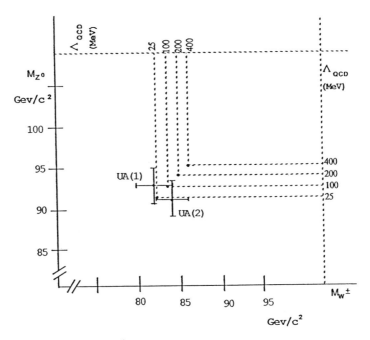

Fig. 1   Values of W and Z^0 masses obtained using Λ_{QCD} and results obtained at CERN.

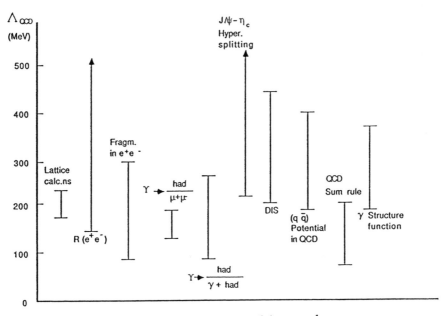

Fig. 2   Summary of Λ_{QCD} values.

gave results with basic differences, in spite of the same environment. These differences were attributed to the energy dependence and other particular properties of the interactions such as the primary particle being a pion or a K–meson, or a proton or a neutron, etc.

A very large amount of data in a vast variety of processes was available. All these data were analysed without a basic notion being considered: i.e. the fact that if an initial state contains a "quark", in the final state this quark possesses a "leading" role.

In order to discover if universality features exist, it is imperative to study each reaction with great care. Each one, singularly. If you do this, you discover the following: no matter if the interaction is

weak,
or **electromagnetic,**
or **strong:**

if in the initial state there is a quark, you will find that in the final state this quark has a "leading" effect. Interaction by interaction, each one taken singularly, this effect has to be subtracted, before the analysis is performed, and the data classified as results obtained in weak, electromagnetic or strong interactions.

Let me repeat, once again: no matter if the interaction is weak, electromagnetic, or strong, whenever there is a leading hadron, its effects have to be subtracted. If you do not do this, you will not find universality features in non–perturbative QCD.

If the leading effects are correctly taken into account, then you will discover that the multi–hadronic states show universality features: no matter if these states are mediated by weak, electromagnetic or strong signals: (W, Z), (photons), (gluons), respectively.

There is a large number of non–perturbative properties. For example, hadron multiplicities, transverse momentum distributions, inclusive hadron energy distributions, and two particle correlations.

2. PROOF OF UNIVERSALITY FEATURES IN NON–PERTURBATIVE QCD

The proof that universality holds true at the non–perturbative level is the content of this lecture.

First, let me start by showing some simple kinematic diagrams for the identification of the correct variables.

2.1 The identification of the correct variables

The diagrams illustrating the three basic examples of strong (fig. 3), weak (fig. 4), and electromagnetic (fig. 5) interactions are given below.

Let us start with the "strong" case.

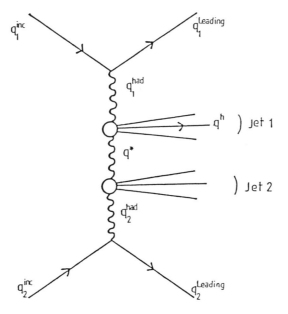

Fig. 3   Schematic diagram for the (pp) interactions.

q_1^{inc} is the four momentum of the incident hadron. It can be a proton, a neutron, a pion, a K–meson etc.

q_2^{inc} is the four momentum of the other incident hadron. As above, it can be any hadron.

$q_1^{leading}$ is the four momentum of the leading hadron. This can be identical to the incident one, or can be different. The more it is different, the less its "leading" effect is going to be. For example if the incident hadron is a proton, and the outgoing leading hadron is another proton, the leading effect is maximum. If the outgoing hadron is a neutron, the leading effect is less pronounced. In fact, from the initial state to the final one, the number of quarks carried is three, in the case of the proton, but only two, in the case of the neutron. And if the final leading hadron is a $\sum^-$, the number of quarks carried through is only one: the leading effect is still less pronounced.

$q_2^{leading}$ is the four momentum of the other leading hadron. As above, the same remarks hold true.

q_1^{HAD} is the four momentum emitted by hadron no. 1, in its transition form q_1^{inc}, to $q_1^{leading}$.

q_2^{HAD} is the four momentum emitted by hadron no. 2, in its transition form q_2^{inc} to $q_2^{leading}$.

$q_{1,2}^*$ is the four momentum exchanged between the two vertices 1 and 2.

q^h is the four momentum of the hadron contained in the multi–hadronic system no. 1, called $(jet)_1$.

$(jet)_1$ is the multi–hadronic system no.1.

$(jet)_2$ is the multi–hadronic system no. 2.

The reader can be suspicious and ask the question: how do you know that this diagram and the corresponding quantities, will suffice to describe the incredibly complicated hadron–hadron collision. The answer is very simple. The diagram is the simplest that can be drawn. And it refers to no theoretical model at all. It represents the simplest kinematic set of variables you can select in order to attempt an individual, reaction by reaction, study of a hadron–hadron collision.

In order to give an example with a specific case, let us consider as (hadron–hadron) collision, the (pp) case. This is in fact the process which has allowed us to discover the universality features mentioned above. We have studied an order of 10^5 (pp) collisions, one by one, using data taken at the CERN–ISR Collider (Ref. 4).

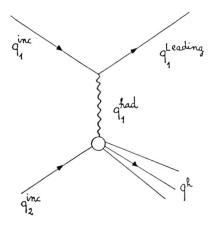

Fig. 4   Schematic diagram for DIS processes.

As it is well known at ISR, or at any other Collider, since ever, the quantity $\sqrt{s} = \sqrt{(q_1^{inc} + q_2^{inc})^2}$ was considered to be the total energy available in the centre of mass system. As our study show this is not true. The quantity $\sqrt{s}$ is to be considered the "nominal" not the "effective" value for the total energy available. In fact, if you examine the final state of a (pp) interaction, you find two leading particles $q_1^{leading}$ and $q_2^{leading}$. On the average, they carry 50% of "nominal" energy. So if you want to analyze this interaction, you have to use $(q_1^{inc} - q_1^{leading})$ to calculate the quantity q_1^{had}. Then you do the same for q_2^{had}. It is $\sqrt{(q_1^{had} + q_2^{had})^2}$ which is the effective total energy available in the interaction. This quantity corresponds to $\sqrt{s}$ in (e^+e^-) annihilation.

Let us see how, in our notation, the classical "Deep Inelastic Scattering" is described. Notice that DIS can be an example of weak or electromagnetic interaction production process: think of $\nu p \to \nu +$ anything and $e^- p \to e^- +$ anything. The diagram illustrating DIS is shown in fig. 4, together with the correspondance between our notation and the classical one.

$$(W)_{DIS} = \sqrt{(q_1^{had} + q_2^{inc})^2} \qquad\qquad (Z)_{DIS} = \frac{q^h \cdot q_2^{inc}}{q_1^{had} \cdot q_2^{inc}}$$

We finally consider the $(e^+ e^-)$ annihilation case (fig. 5)

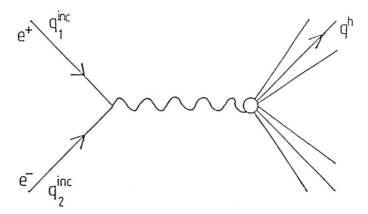

Fig. 5   Schematic diagram for the $(e^+ e^-)$ annihilation.

$$(\sqrt{s})_{(e^+ e^-)} = \sqrt{(q_1^{inc} + q_2^{inc})^2} = \sqrt{(q_1^{had} + q_2^{had})^2} = \sqrt{(q_{tot}^{had})^2}$$

Notice:

$$q_1^{inc} = q_1^{had}$$
$$q_2^{inc} = q_2^{had}$$

$$(x_R)_{e^+ e^-} = \frac{2 q^h \cdot q_{total}^{had}}{q_{total}^{had} \cdot q_{total}^{had}} = \frac{2 E^h}{\sqrt{s}}$$

From what precedes it is clear that, in (pp) interactions (see fig. 3) you can work out the right quantities to conform to DIS or (e^+e^-). For clarity we list them.

(e^+e^-)(LIKE) QUANTITIES

$$\sqrt{(q_{total}^{had})^2} = \sqrt{(q_1^{had} + q_2^{had})^2} = (\sqrt{s})_{pp}^{had}$$

$$(x_R)_{pp}^{had} = \frac{2q^h \cdot q_{total}^{had}}{q_{total}^{had} \cdot q_{total}^{had}}$$

DIS (LIKE) QUANTITIES

$$(W^2)_{pp}^{had} = (q_1^{had} + q_2^{inc})^2$$

$$(Z)_{pp}^{had} = \frac{q^h \cdot q_2^{inc}}{q_1^{had} \cdot q_2^{inc}}$$

In the e^+e^- system the total energy available in the centre of mass system is $\sqrt{(q_1^{inc} + q_2^{inc})^2}$, but for DIS the total available energy in the centre of mass is $(W)_{DIS}$ as defined above.

2.2 Experimental results

Our method means to put all the reactions, be they of the (hadron–hadron) type such as (pp) or (hadron–lepton) such as DIS or (lepton–lepton) such as (e^+e^-), on the same mathematical basis. And then to study if universality features show up.

To achieve this aim we have considered the following eight quantities;

– the fractional energy distribution $\frac{d\sigma}{dx}$;
– the average number of charged particles $< n_{ch} >$;
– the ratio of the average energy to the total energy in the charged channel: $< E_{ch} > /E_{total}$;
– the transverse momentum distributions $d\sigma/dp_t^2$;
– the normalized transverse momentum distribution $\left(\frac{d\sigma}{dp_t/<p_t>}\right)$;
– the event planarity;
– the two–particle correlations;
– the scale–breaking effects.

I will now take up these features one at a time.

• The first is the fractional energy distribution, which is shown in fig. 6. Notice that this pp interaction occurred at the ISR with a nominal energy of 62 GeV. But when you have 31 GeV on 31 GeV as mentioned earlier, it is not true that the total "effective" energy is 62 GeV. Instead it is 62 GeV minus the amount taken by the leading particles. So this high energy interaction can be analysed in terms of the effective energy available for particle production and this can be very low, as for example a few GeV, to be compared with e^+e^- data. We have taken SLAC data at the same energy, and we find the remarkable result that the fractional energy distribution is indeed the same.

Next you can consider those events where the leading particles have carried away less energy, thereby making the effective available energy higher. You can compare this with (e^+e^-) and you will again find remarkable agreement (see figs. 7 and 8). We now change the nominal primary energy of the hadronic system. This is shown in fig. 9. The pp data were taken at the nominal energies 62, 44, and 30 GeV, but with similar effective energies in the range 10–16 GeV. The e^+e^- data from TASSO are also shown. Figures 10 and 11 show similar results, but for higher effective ISR energies. These data show that the (pp) interactions at ISR "simulate" multi–hadronic production equivalent to e^+e^- in the 30 GeV energy range.

• Figure 12 shows the average number of charged particles produced in pp and e^+e^- collisions. If we analyse the (pp) data as other groups have done in the past, then we get excellent agreement with previous (pp) results, but total lack of similarity with (e^+e^-) data. If we analyse the same (pp) data in terms of the effective energy available, by subtracting the leading protons, then we jump onto the (e^+e^-) curve. This means that, regardless of whether you produce particles starting from e^+e^- or pp, the average number of charged particles is the same.

• Now I move on to examine another quantity, the ratio of the charged energy over total energy, which tells you how much energy in a reaction is given to charged particles compared to the total energy. If you do not use the method of subtracting leading proton effects, there is no denominator because the effective E_{total} is unknown in the (pp) case. But if you subtract the effects of the two leading protons, then you know how much energy is really available, and you have the denominator. The results of this are given in fig. 13. All the data available show that, in spite of the different effective energies, $< E_{charged} > / < E_{total} >$ remains unchanged. And this is in excellent agreement with e^+e^- data.

• Next we consider p_t^2 distributions. Fig. 14 shows the two domains: high p_T jets and low p_T jets. The Capital T indicates that the transverse momentum is measured with respect to the incident beam. While p_t (small t) momentum is measured with respect to the jet axis (see fig. 15).

Figure 16 shows the remarkable result that p_t^2 distributions in (pp) interactions at nominal energies of 30 GeV, but at effective energies of a few GeV, look the same

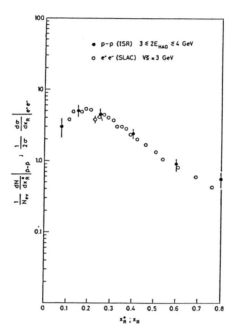

Fig. 6   The inclusive single–particle fractional momentum distributions in the interval

3 GeV $\leq 2E_{HAD} \leq$ 4 GeV obtained from data at $(\sqrt{s})_{pp} = 30$ GeV. Also shown are data from MARK I at SPEAR See text for definition of variables. Notice that the "effective energy" is here given in terms of a quantity $2E_{HAD} \simeq \sqrt{(q_{tot}^{had})^2}$ (see references 3 and 4).

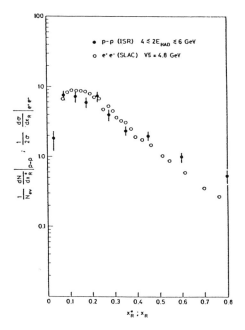

Fig. 7   As fig. 6 but in the interval 4
GeV ≤ $2E_{HAD}$ ≤ 6 GeV.

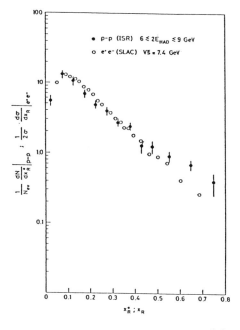

Fig. 8   As fig. 6 but in the interval 6
GeV ≤ $2E_{HAD}$ ≤ 9 GeV.

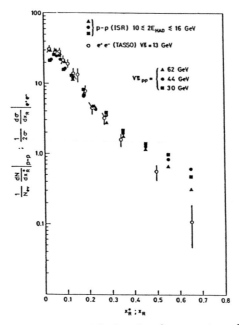

Fig. 9   The inclusive single–particle fractional momentum distributions in the interval
10 GeV $\leq 2E_{HAD} \leq$ 16 GeV obtained from data at different $(\sqrt{s})_{pp}$. Also shown are data from TASSO and PETRA.

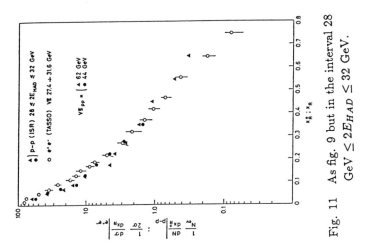

Fig. 11   As fig. 9 but in the interval 28
GeV ≤ 2E_{HAD} ≤ 32 GeV.

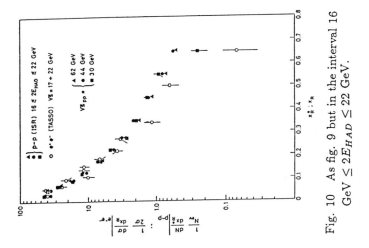

Fig. 10   As fig. 9 but in the interval 16
GeV ≤ 2E_{HAD} ≤ 22 GeV.

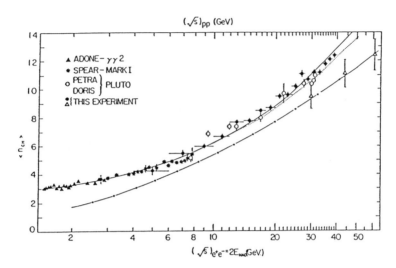

Fig. 12 Mean charged–particle multiplicity (averaged over different $(\sqrt{s}_{pp})$) vs. $2E_{HAD}$, compared with (e^+e^-) data. The continuous line is the best fit to our data according to the formula $<n_{ch}> = a+b\,exp\,\{C\sqrt{\ell n(s/\Lambda)^2}\}$. The dotted line is the best fit using PLUTO data. The dash–dotted line is the standard (pp) total charged particle multiplicity with, superimposed, our data as open triangular points.

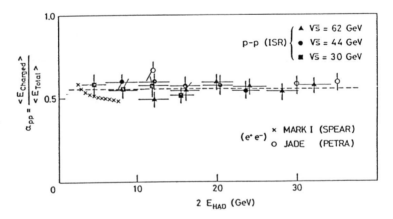

Fig. 13 The charged–to–total energy ratio α_{pp}, obtained in (pp) collisions plotted vs. $2E_{HAD}$ and compared with (e^+e^-) obtained at SPEAR and PETRA.

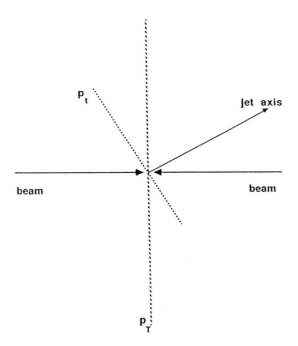

Notice that the p_t and p_T planes are both perpendicular to the figure

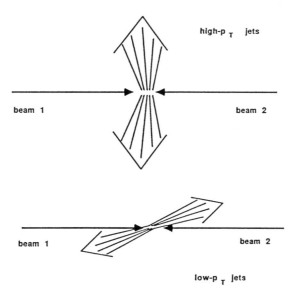

Fig. 14   The two p_T domains: high p_T jets and low p_T jets.

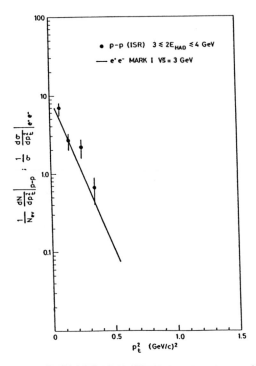

Fig. 16   The inclusive single–particle transverse momentum distribution in the interval
3 GeV $\leq 2E_{HAD} \leq 4$ GeV obtained from data at $(\sqrt{s})_{pp} = 30$ GeV.
Also shown is the fit to the (e^+e^-) data obtained at SPEAR (continuous line).

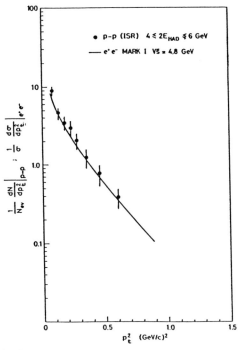

Fig. 17   As in fig. 16 but in the interval 4 GeV $\leq 2E_{HAD} \leq$ 6 GeV.

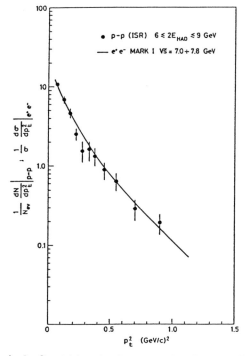

Fig. 18   As in fig. 16 but in the interval 6 GeV $\leq 2E_{HAD} \leq$ 9 GeV.

when compared to e^+e^- data at 3 GeV. Figs. 17 and 18 show similar results but at different effective energies, again the agreement is very good. When we go up even higher in energy we start to see some differences (figs. 19 and 20), which we attribute to problems with the apparatus; high effective energy means low leading effects, so you start getting contamination from secondary particles.

- The deviations seem to get worse at high p_t. However, there is a way to reduce sistematic effects. Namely: the use of a new variable, the transverse momentum over the average transverse momentum of the jet. Once this is done a remarkable universality curve is found. These results are shown in fig. 21. This effect can be better understood by considering the event planarity.

- The event planarity can be visualized as follows. Imagine a jet perpendicular to the plane of the diagram in fig. 22. In the standard hydronization plus a gluon emission, we define two planes, the in–plane and the out–plane, as shown in the diagram. Using the definitions in table 1, fig. 23 shows that the transverse momentum in the in–plane is higher than expected. Now we look carefully at the p_t–out and p_t–in distributions. We see that there is not an excellent agreement with the e^+e^- data. This is due to the fact that e^+e^- in the out–distribution have higher p_t values. But these high p_t values, which are also reflected in the in–distribution, are such that when you normalize the values to the average, you get the same result. Figure 24 shows the results for another energy range.

- The definition of the two–particle correlation function is given in table 2. It has often been stated that the particle correlations in multi–hadronic systems produced in e^+e^- and in hadronic collisions are not the same. The reason for this is that the leading effects were not subtracted. The graphs in figs. 25 show the correlation functions measured in our experiment and that measured in previous experiments which were claiming differences with e^+e^- annihilation data. These data show that when we do the analysis in the standard way, without subtracting leading effects, our data are exactly the same as the data from previous experiments. But when we apply the corrections (as shown in fig. 26), we find that our data do agree quite well with the e^+e^- data. We are also able to calculate the dependence of the correlation function on the effective hadronic energy (fig. 27). We cannot compare this result with (e^+e^-) because in the (e^+e^-) case this analysis has not been made. However, the two curves in the picture compare the results obtained by doing the subtraction and by not doing the subtraction.

- The next topic is that of scale–breaking effects. Scale breaking effects are not expected in simple quark–parton models. They are predicted to exist by QCD as a consequence of the emission of gluons from the primary quarks. These events have been looked for in the multi–particle hadronic states produced in e^+e^- annihilation. In terms of quantitative QCD predictions, gluon emission leads to a depletion of particles at large x_R^* and an increase at small x_R^* with increasing effective energy.

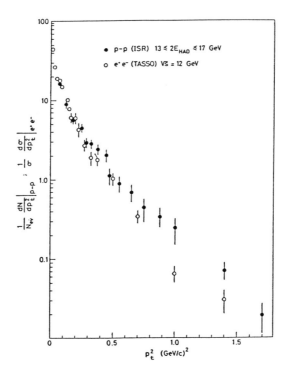

Fig. 19 The inclusive single–particle transverse momentum distributions in the interval
11 GeV $\leq 2E_{HAD} \leq$ 13 GeV obtained from data at $\sqrt{s} = 62$ GeV. Also shown are data from (e^+e^-) annihilation obtained by TASSO at PETRA.

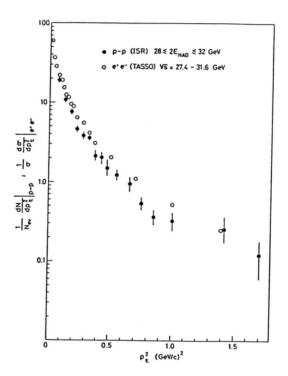

Fig. 20   As in fig. 19 but in the interval 28 GeV $\leq 2E_{HAD} \leq$ 32 GeV.

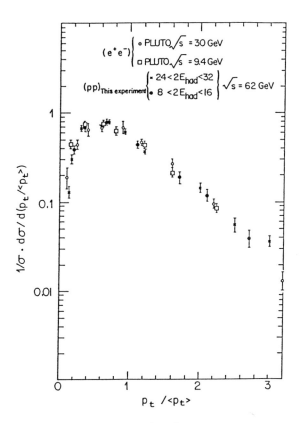

Fig. 21 The differential cross-section $\frac{1}{\sigma} \frac{d\sigma}{d\frac{p_t}{<p_t>}}$ vs. the reduced variable $p_t/ <$ $p_t >$. These distributions allow a comparison of the multiparticle systems produced in (e^+e^-) annihilation and in (pp) interactions in terms of the renormalized transverse-momentum properties.

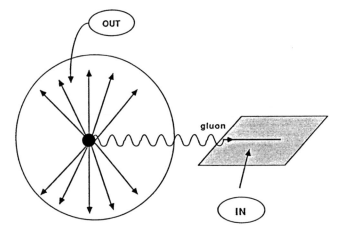

"JET" AXIS PERPENDICULAR TO THE PLANE OF THE FIGURE.
The "OUT" plane is the plane of the figure. The "IN" plane
contains the "gluon".

Fig. 22 Definition of the in–plane and the out–plane in the standard hadronization plus a gluon emission.

Table 1

DEFINITION OF EVENT PLANARITY

The planarity of the events has been studied by means of the $< p_t >^2_{in,out}-$ distributions.

These quantities are determined in the following way:

1. Construct Tensor

$$M_{\alpha\beta} = \sum_{j=1}^{N} P_{j\alpha} \cdot P_{j\beta} \ (\alpha, \beta = 1, 2)$$

where $N \equiv$ number of particles in the event.

2. Determine the eigenvectors $\vec{n}_1, \vec{n}_2$ and the eigenvalues $\Lambda_1, \Lambda_2 (\Lambda_1 < \Lambda_2)$.

3. Define

$$< p_t^2 >_{OUT} = \frac{\Lambda_1}{N} = \sum_{j=1}^{N} (\vec{P}_j \cdot \vec{n}_1)^2$$

$$< p_t^2 >_{IN} = \frac{\Lambda_2}{N} = \sum_{j=1}^{N} (\vec{P}_j \cdot \vec{n}_2)^2$$

Table 2

DEFINITION OF TWO PARTICLE CORRELATION FUNCTION

The two particle correlation function in rapidity space $R(y, y')$ is defined as

$$R(y, y') = \frac{\rho_2(y, y')}{f \rho_1(y) \rho_1(y')} - 1$$

where

$\rho_1(y)$ is the normalized single–particle inclusive rapidity distribution:

$$\rho_1(y) = \frac{1}{\sigma_{in}} \frac{d\sigma}{dy}$$

$(\sigma_{in} = $ inelastic cross–section$)$

$\rho_2(y, y')$ is the two–particle normalized rapidity distribution

$$\rho_2(y, y') = \frac{1}{\sigma_{in}} \frac{d\sigma(y, y')}{dy \, dy'}$$

f is a normalization factor:

$$f = \frac{<n_{ch}(n_{ch} - 1)>}{<n_{ch}>^2}$$

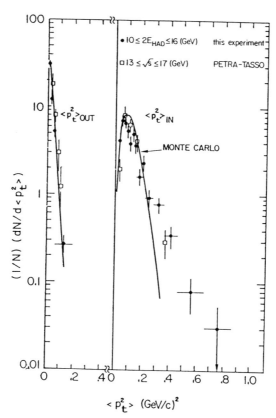

Fig. 23 The average transverse–momentum squared distributions for "out" and "in" cases, in the $(10 \div 17 \text{ GeV})$ "effective energy" range.

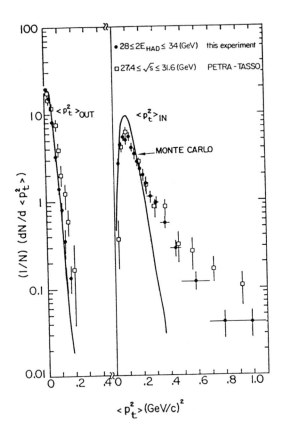

Fig. 24   As in fig. 23 but for (27 ÷ 34 GeV) "effective energy" range.

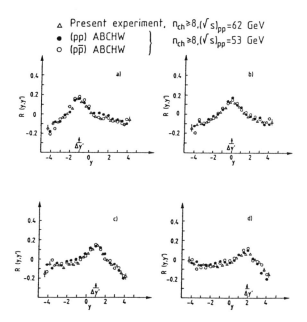

Fig. 25 Two–particle correlations in rapidity space $R(y, y')$ for different y' intervals [a) -1.75 $\leq y' \leq -0.75$, b) $-0.25 \leq y' \leq 0.25$, c) $0.75 \leq y' \leq 1.25$,
d) $1.75 \leq y' \leq 2.25$] as measured by the ACBHW Collaboration at
$(\sqrt{s})_{pp} = 53$ GeV in (pp) (black points) and $(p\bar{p})$ (open circles) collisions and by the present experiment (open triangles) at $(\sqrt{s})_{pp} =$
62 GeV in (pp) collisions. These results are obtained from minimum–
bias events without leading–proton subtraction for an observed charged
multiplicity: $n_{ch} \geq 8$.

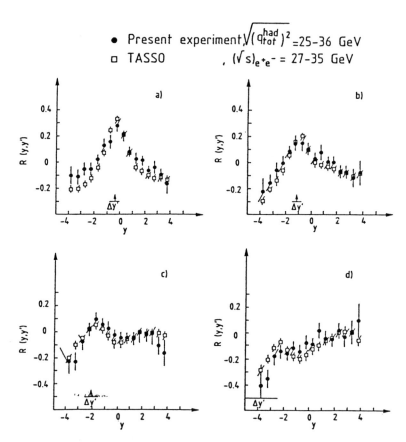

Fig. 26 The two particle in rapidity space $R(y, y')$ for different y' intervals, as measured in (pp) interactions (present experiment) after the leading proton subtraction in the $\sqrt{(q_{tot}^{had})^2}$ range 25 GeV to 36 GeV (black points) compared with results obtained in (e^+e^-) annihilation by the TASSO Collaboration at $(\sqrt{s})_{e^+e^-}$ between 25 GeV and 35 GeV (open squares).

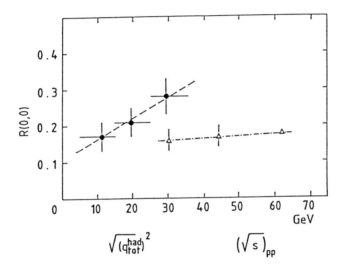

Fig. 27 Correlation function $R(y, y')$ measured at $y = y' = 0$, vs. $\sqrt{(q_{tot}^{had})^2}$. The data are indicated as black points. The broken line is the best fit. For comparison, the open triangles show $R(0,0)$ vs. $(\sqrt{s})_{pp}$. In this case the analysis of the final state is made without subtracting the leading proton effects. The dash–dotted line is the best fit to these data. Notice that the abscissa for the black points is $\sqrt{(q_{tot}^{had})^2}$; for the open triangles, it is $(\sqrt{s})_{pp}$.

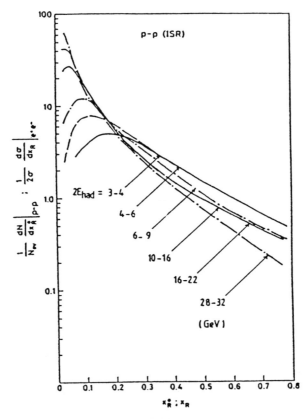

Fig. 28   Fits to the distributions $\frac{1}{N_{\mathrm{ev}}} \frac{dN}{dx_R^*}$ as measured at the ISR ((pp) collisions) for different values of the "effective energy". Here the values of
the "effective energy" are given by the quantity $2E_{HAD} \simeq \sqrt{(q_{tot}^{had})^2}$.
The relevant point to notice here is that all these data have been obtained using the same "nominal" total energy of $\sqrt{s} = 62$ GeV.

If you apply the leading particle correction, you can see the effect in the results of fig. 28. Notice that all these curves are obtained from (pp) interactions at the same "nominal" total energy $(\sqrt{s})_{pp} = 62$ GeV. If you sum up all these data you get the standard $\frac{d\sigma}{dx}$ distributions for (pp) interactions where the correlation with (e^+e^-) is completely lost, because you are comparing the sum of an enormous number of effective energies as it is in the (pp) case, with one and only one "effective" energy, as it is in the case of (e^+e^-) annihilation. If you analyse 100,000 pp interactions and correct each interaction for the leading protons, you can classify the (pp) interactions in terms of low, high, and very high effective energy. You are then ready to repeat the same type of analysis as e^+e^- people have done for small and increasing x_R^* values. We find (fig. 29) that also in the (pp) case the slope changes according to the expression:

$$\frac{1}{N_{ev}} \frac{dN}{dx_R^*} = A\left[1 + C \, \ell \, n\left(q_{tot}^{had}\right)^2\right]$$

The values of parameter C are plotted as a function of x_R^* in fig. 30. The values of C measured at TASSO are reported. The data show that also in the scale–breaking effects the multiparticle systems produced in pp interactions show striking analogies with the multiparticle systems produced in e^+e^- annihilations.

2.3 The leading effect in (e^+e^-) annihilation

In (e^+e^-) annihilations you do not have leading effects; the produced jets do not have a leading particle in 99% of the cases. But in 1% of the cases you do have leading effects. The standard e^+e^- jets are shown in fig. 31, with jet 1 and jet 2 more or less back–to–back, and in about 1% of the cases one jet with a leading particle. For example, D^* production and heavy quark production give rise to leading effects. An important question is what effective energy is available in the system. Should you use the $\sqrt{s}$ at e^+e^- or should you use $\sqrt{s - (leading\ particle)}$? The answer is given in fig. 32, which shows $\frac{d\sigma}{dX}$ for TASSO data. You can see the difference between the data with D^*–jets, and those with "average jet". When the TASSO group took data at 14 GeV (fig. 33), the agreement became very good (Ref. 5). In other words, to use 14 GeV data corresponds to subtracting the leading particle. Furthermore, if you compare the TASSO data where the leading particle has been subtracted with pp data where the effective energy is comparable with the TASSO data, you find perfect agreement (fig. 34). No matter how you produce these multiparticle systems, the important point is that you must be careful in evaluating the correct energy available for particle production. There is something even more dramatic, which is the comparison of the transverse distribution of the particles inside these jets. If you use TASSO data at 14 GeV and our data, you get the results shown in fig. 35. This is a unique case where proton collisions compete with e^+e^- collisions. Why? Because in (e^+e^-) the leading effects are at the 1% level and therefore the errors are very large. However, the results are in excellent agreement.

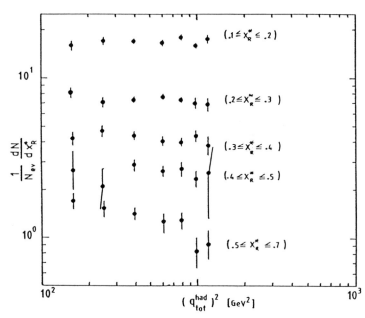

Fig. 29   The quantity $\frac{1}{N_{\text{ev}}} \cdot \frac{dN}{dx_R^*}$ plotted vs. $(q_{tot}^{had})^2$ for different ranges of x_R^*.

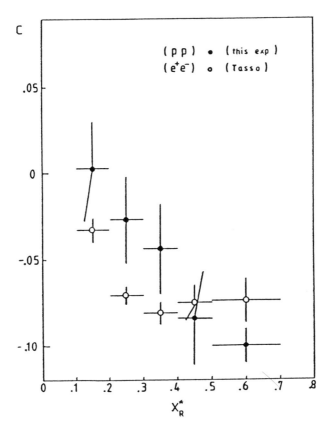

Fig. 30   The values of the coefficient C vs. x_R^* as measured in (pp) interaction
(this
experiment) and in e^+e^- annihilation by TASSO Collaboration.

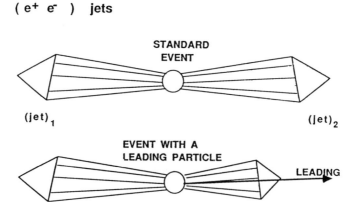

Fig. 31   Standard e^+e^- jets: standard event and event with a leading particle.

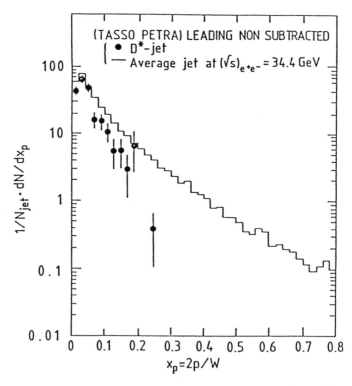

Fig. 32   PETRA–TASSO Collaboration. The distribution $\frac{1}{N_{jet}} \cdot \frac{dN}{dx_p}$ measured in standard jets and in jets containing a D^* at $(\sqrt{s})_{e^+e^-} = 34.4$ GeV.

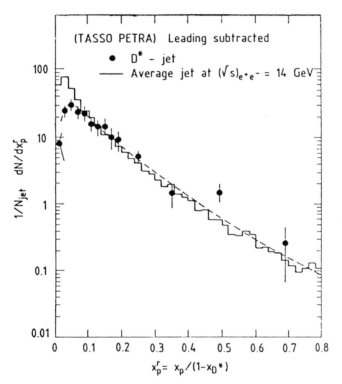

Fig. 33   As fig 32 but at $(\sqrt{s})_{e^+e^-} = 14$ GeV.

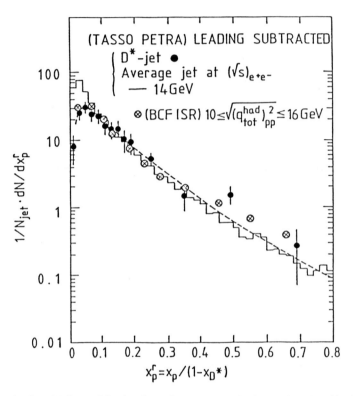

Fig. 34 As fig. 32 but with the data from our analysis at the ISR ((pp) interactions) with "effective energy" in the $10 \div 16$ GeV range.

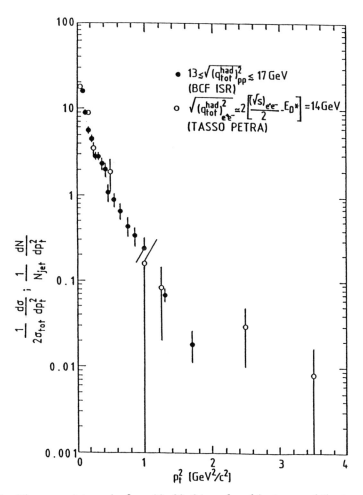

Fig. 35 The same jets as in figs. 32, 33, 34 analysed in terms of the transverse
p_t momentum distribution.

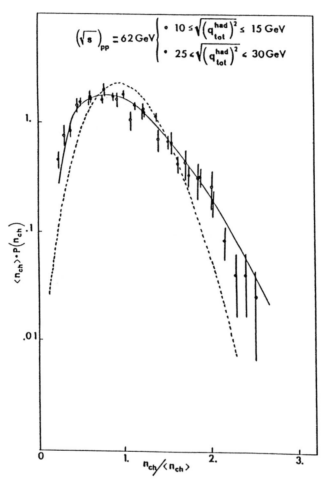

Fig. 36 The multiplicity distributions measured in (pp) interactions for two different values of $\sqrt{(q_{tot}^{had})^2}$ (open circles: 10 GeV $\leq \sqrt{(q_{tot}^{had})^2} \leq 15$ GeV; full circles: 25 GeV $\leq \sqrt{(q_{tot}^{had})^2} \leq 30$ GeV). The solid curve is the best fit to (pp) data. The nominal (pp) c.m. energy was $(\sqrt{s})_{pp} = 62$ GeV. The dashed line is the best fit to (e^+e^-) data.

2.4 Where the difference

The next question is whether there is any place where some difference can show up. We were looking for possible sources of difference, and we found that there is something in the multiplicity distributions. In fig. 36, we plot the multiplicity distributions as a function of $n_{ch}/ < n_{ch} >$ using e^+e^- and pp data. We observe that the averages are the same, which is why the curves of $< n_{ch} >$ versus energy are the same. But the distribution functions are not the same; probably because in the (pp) case you have quark fragmentation due to the fact that we are studying low p_T jets. In the other case, (e^+e^-), you have quark and gluon fagmentation. But I do not want to enter into an analysis which is outside our experimental results.

Let me just consider an application of this finding. Suppose you want to predict the probability distribution for the number of charged particles, $P(n_{ch})$, produced in a multi–Tev Collider. Thanks to UA(5) there are data available for multiplicity distributions in (pp) interactions at $\sqrt{s} = 540$ GeV (Ref. 6). We have assumed that the function $P(n_{ch})$ (fig. 36) scales with energy. Next we have assumed for (pp) interactions at 540 GeV the same leading effects, as measured at the ISR. Then we have evaluated the expected $P(n_{ch})$: this is shown by the full–line in fig. 37. Had we taken for $P(n_{ch})$ the value measured in (e^+e^-) (fig. 36), the predictions at $\sqrt{s} = 540$ for (pp) interactions would be given by the dotted line. The conclusion is that the UA(5) data support the "leading effect" being present at Collider energy as it is at ISR energies. Moreover the $P(n_{ch})$ follows the (pp) and not the (e^+e^+) finding. Thus we have found something where a difference emerges between properties of multi–hadronic systems produced in (e^+e^-) annihilation and in (pp) interactions. However for a direct comparison of this finding, we would need (e^+e^-) data at energies suitable for our (pp) range of "effective" energies.

2.5 Compare (pp) with DIS: high and low p_T physics

Let me review the situation for deep inelastic scattering. The important quantity in this case is:

$$W^2 = \left(q_1^{had} + q_{proton}^{inc} \right)^2$$

The $< n_{ch} >$ distribution is shown in fig. 38 for pp and νp interactions. Notice that the two sets of data do not overlap, this is because the pp energies (ISR) are higher than the highest neutrino energies available for DIS. Fig. 39 shows the fractional energy distribution for pp and DIS, using data from the EMC.

What conclusions can we draw? The results presented here for low p_T, pp interactions, used $\sqrt{(q_{tot}^{had})^2}$ for the comparison with e^+e^- data and $\sqrt{(q_1^{had} + q_2^{inc})^2}$ for the comparison with DIS data. The conclusion is: it is not true that, in order to compare (pp) interactions with e^+e^- and DIS processes you need to select high p_T interactions.

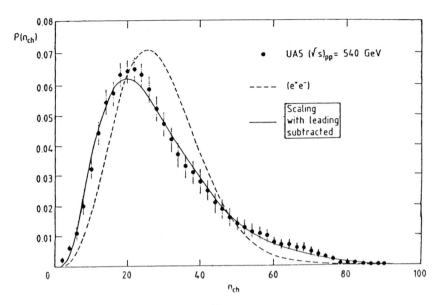

Fig. 37 Multiplicity distribution at $(\sqrt{s})_{pp} = 540$ GeV. Points are from UA5 Collaboration. The full line shows the expected $P(n_{ch})$ at $(\sqrt{s})_{pp} = 540$ GeV assuming that $P(n_{ch})$ in fig. 36 scales with energy. Also shown is the prediction in the case that the $P(n_{ch})$ from e^+e^- had been used as input (dotted line).

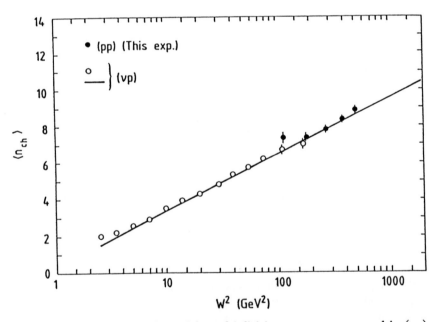

Fig. 38 The average charged–particle multiplicities $< n_{ch} >$ measured in (pp) at $(\sqrt{s})_{pp} = 30$ GeV, using DIS–like analysis, are plotted vs. W^2 (black points). The open points are the (νp) data and the continuous line is their best fit.

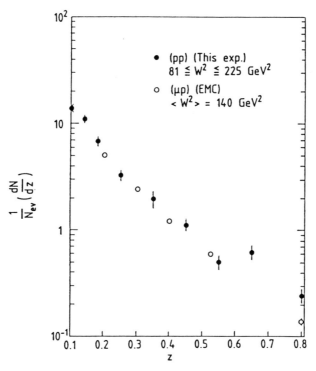

Fig. 39 The inclusive distribution of the fractional energy z for (pp) interactions in the energy interval $81 \text{ GeV}^2 \leq W^2 \leq 225 \text{ GeV}^2$ compared with the data from (μp) reactions at $< W^2 > = 140$ GeV.

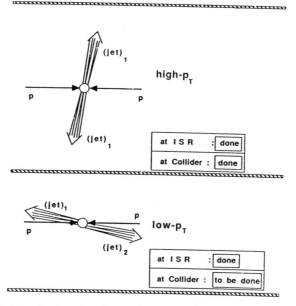

Fig. 40 Synthesis of what has been done and what needs to be done in the study of multi–hadronic systems at the ISR and at the Collider.

The data reported in this review lecture show that, low p_T proton–proton in-teractions, reproduce the basic features of the multi–hadronic systems produced i (DIS) and e^+e^- interactions. The key point is not the selection of high p_T processe but the subtraction of the leading effects.

2.6 Extrapolation to Collider physics

Now I move on to the very exciting comparisons we made with Collider phys. The extrapolation of our method to the CERN $\bar{p}p$ Collider is possible because th are two ways of determining $\sqrt{(q_{tot}^{had})^2}$: one using the two leading particles and other, when the leading method is out of experimental reach, using the measureme of the total energy carried by the two back–to–back jets. Again, the basic quant in comparing multihadronic states is the effective energy available for particle p duction, not the nominal one. In the case of two back–to–back jets, this is giv by:

$$\sqrt{\left(q_{tot}^{had}\right)^2} = \left(\sum_{i=1}^{n} E_i\right)_{jet1} + \left(\sum_{i=1}^{m} E_i\right)_{jet2}$$

where E_i is the energy of particle "i" in each of the two back–to–back jets wit multiplicities n for jet 1 and m for jet 2.

There is an important experimental difference between the two ways of determin ing $\sqrt{(q_{tot}^{had})^2}$. The leading method is straightforward; it only needs the measurement of the two leading particles. The other method requires the measurement of all th particles in the two jets, including the neutral ones. The important point here is tha the remarkably unique range of effective hadronic energies available at the Collider allows us to make a large jump from ISR energies, and to produce clear evidence fo or against our expectations based on the results obtained at the ISR.

For the two leading particle method you have to calculate the quantity:

$$\sqrt{\left(q_{tot}^{had}\right)^2} = \sqrt{\left(q_1^{inc} + q_2^{inc} - q_1^{lead} - q_2^{lead}\right)^2}$$

so you must know the two leading particles. All of this can be done for low p_T and high p_T.

What has been done so far is illustrated in fig. 40.

The low p_T physics at the Collider is open for competition.

Now we can ask what kind of extrapolations can be made in going from ISR to Collider.

The leading proton effect measured at the ISR is shown in fig. 41. It appears to be energy independent so we can assume the validity of this curve up to Collider energies and use it as an input. The second input is the number of charged particles as a function of the effective energy, shown in fig. 42. The third input is the multiplicity distribution plotted in fig. 36. To sum up, the basic input data are:

(1) $\frac{d\sigma}{dX_F}$ from ISR, (fig. 41);
(2) $<n_{ch}>$ versus (q_{tot}^{had}) from ISR or $\sqrt{s}$ from e^+e^- (fig. 42);
(3) Multiplicity distributions from ISR (fig. 36). With these three inputs it is

possible to predict the average number of charged particles at Collider energies, the forward–backward correlation functions, and the multiplicity distribution.

- Figure 43 shows our prediction for the average number of charged particles. The data measured by UA(5) are on our curve.

- By forward–backward correlation the following is meant: you study the multiplicity in one hemisphere, (called forward), and the number of particles you see in the other (called backward). It has been known for a very long time that there is a correlation between these two hemispheres. If you have many particles in one hemisphere, then you have many in the other as well. The same, if you have few particles. Our claim is that on the average this well known "Forward–Backward" correlation has nothing to do with long range effects. If you detect many particles in one hemisphere, then you are at high (q_{tot}^{had}), and therefore you are bound to find many particles in the other hemisphere. If you pick up only a few particles in one hemisphere, you are picking up a low effective energy interaction, this is why you see a few particles in the other hemisphere, as well. The proof of this is presented in fig. 44. Here the world's data on the correlation function are shown. You see that the straight line has a logarithmic dependence on s and that the Collider data agree well with our extrapolation. We conclude from this that what has been observed, in the past years all over the low energy range and now at the Collider, has nothing to do with long–range correlation. It has to do with the fact that also at the Collider the leading proton effects are present.

- Now I want to tell you about a very hot result, the most recent one. It refers to $P(n_{ch})$ measured at the Collider. We can make a firm prediction, using the same three inputs mentioned above. Our prediction for $P(n_{ch})$ at 540 GeV is shown in fig. 45, together with the UA(5) data. The agreement is remarkable.

Let me add a further remark on $P(n_{ch})$. It has been recently stated that if you take the charged particle distributions at 14, 62, and 540 GeV (shown in fig. 45), you will see the effects of scale–breaking. Our claim is that these data have nothing to do with scale–breaking. In fact we are able to predict these data without any need of scale–breaking, as the superimposed curves resulting from our calculations show.

3. CONCLUSION

The final conclusion is: no matter if the interaction is strong, electromagnetic, or weak, the basic quantity which allows to discover the existence of universality features in the multihadronic states is the relativistic invariant quantity (q_{tot}^{had}). These universality features hold in a very crucial domain of QCD phenomena: the non–perturbative one.

It is appropriate to conclude with a quote from Amhed Ali, who is a great specialist in perturbative QCD. He says:

That such universality holds, is by no means obvious or trivial. The experimental confirmation of this universality is an important achievement of experimental high energy physics. The fact that this feature is not yet reproducible in theory from first principles is just a reflection of the weakness of the present theoretical apparatus.

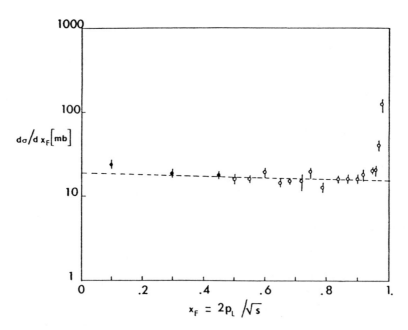

Fig. 41 Best fit to the leading proton effect.

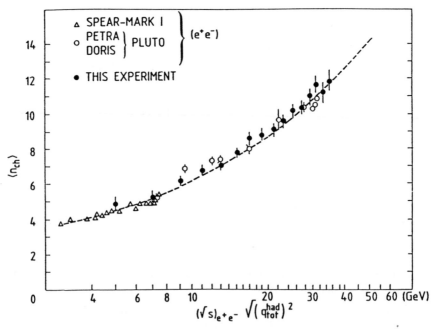

Fig. 42 Mean charged multiplicity as measured in e^+e^- and pp (using the leading proton subtraction).

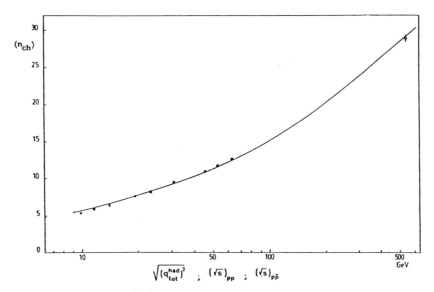

Fig. 43 The predicted (pp) mean charged multiplicities (full curve). Also shown are the experimental measurements. Notice the UA(5) value: the highest energy point, which is in excellent agreement with our predictions.

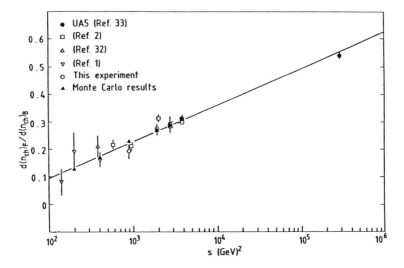

Fig. 44 Comparison of our prediction with the "correlation–strength" $\alpha = d < n_{ch} >_F / d(n_{ch})_B$ measured, in (pp) interactions as a function of s. The full line is the best fit of our Monte Carlo results.

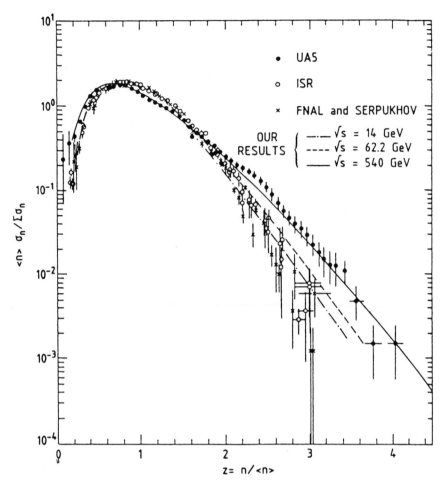

Fig. 45 Distribution of charged particle multiplicities at various energies. Our predictions are in excellent agreement with the data. Notice that in our predictions there is no need for scale-breaking effect. The basic ingredient being the leading effect.

References

1) A. Ali, private communication.

2) See for example: Horst D. Wahl, "Review of weak boson production in antiproton–proton collisions", Proceedings of the XV International Symposium on Multiparticle Dynamics, Lund, Sweden, 11–16 June, 1984, 410.

3) M. Basile et al., Physics Lett. **B132**, (1980) 367.
   M. Basile et al., Nuovo Cimento **58A**, (1980) 193.
   M. Basile et al., Nuovo Cimento **66A**, (1981) 129.

4) M. Basile et al., Nuovo Cimento **79A**, (1984) 1 and references therein.

5) TASSO Collaboration, DESY 83–114.

6) UA5 Collaboration, K. Alpgard et al., Physics Lett. **121B** (1983) 209.

DISCUSSION

CHAIRMAN: PROF. A. ZICHICHI

SCIENTIFIC SECRETARY: R. BLUHM

CHAO:

I would like to ask two questions regarding the differences between e^+e^- and h–h processes. The multiplicity distributions for these processes are different even after subtracting the leading-particle effect, but the p$\bar{\text{p}}$ events you chose for comparison were low-p_T events. Have you ever looked at the multiplicity distribution of the high-p_T jet and compared it with the one for e^+e^-? It seems to me that these would be more similar because they are both hard processes.

ZICHICHI:

So far, within experimental errors, the high-p_T and low-p_T multiparticle systems have the same properties. Furthermore, it is not true that $\langle n_{ch} \rangle$ is different for h–h and e^+e^- processes once the leading-particle effects are subtracted. We have not yet made distinct measurements of high-p_T versus low-p_T — we wanted to, but the ISR had closed down. However, seeing all data known so far up to ISR energies, low-p_T (our data) and high-p_T (other people's data) multiparticle systems show degenerate properties. So far nothing has been found, which justifies the old 'myth' that only high-p_T data in h–h interactions can be compared with e^+e^- data.

CHAO:

My second question has to do with long-range correlations. You said this morning that these are due to events having different effective energies. Have you ever analysed the long-range correlations of p$\bar{\text{p}}$ events after fixing the effective energy and compared them with corresponding e^+e^- events?

ZICHICHI:

We have not enough data to do what you are suggesting. The analysis we performed used data from other experiments, which we had to reinterpret using the leading-particle subtraction method. It is in this way that we derived the backward–forward correlation. I would be very surprised if by doing what you are suggesting (which, by the way, we wanted to do) our results should change. We are very concerned about looking at all possibilities. One key check we have carried out, however, was that of working at different ISR energies. At these different energies we picked up the same effective energies, and the results were identical. At the Collider, we could do what you have said, and we hope to be able to do so in the future.

CHEN:

I would like to make a comment on the first question. I believe that in your analysis the multiplicity distribution does not depend on p_T and that the slight differences in the e^+e^- and pp multiplicity distributions are due to the fact that in e^+e^- annihilation the heavy quarks and light quarks are produced equally, whereas in pp the light quarks are predominant. This is most likely to be the cause of the difference.

LLOYD:

Professor Zichichi, I have a question about universality and the distribution of transverse momentum. Suppose you take two bottles of Coca Cola, smash them together, and look at the transverse-momentum distribution of the debris. If we compare this distribution with the p_T distribution for a collision between two high-energy electrons or protons, we will notice many similarities, modulo a scale factor. We will even see jets which, if not gluey, are at least sticky. My question, then, is how much of the fit between the e^+e^- results and the $p\bar{p}$ results were due to just kinematic factors and how much was due to QCD universality?

ZICHICHI:

No, kinematic factors have nothing to do with it. There is no comparison between the classical way of plotting these results and QCD. The analysis with leading subtraction is very significant and is not at all a matter of kinematics. In fact, if you do not use the 'leading' subtraction method, there is no possibility of putting e^+e^-, h–h, and DIS physics on the same basis.

QIAN:

When we do perturbative QCD calculations, we have to choose a certain renormalization subtraction scheme. The popular schemes are MS (minimal subtraction), $\overline{MS}$, MOM (momentum subtraction), $\overline{MOM}$, etc. Corresponding to each scheme we have different values of Λ. Sometimes these Λ's differ from each other by significant amounts. For example, $\Lambda_{\overline{MS}} \approx (2.7) \Lambda_{MS}$. Which renormalization scheme is the experimental Λ related to?

ZICHICHI:

This question you ask does not concern what I have done. I have used what are considered to be the accepted values of Λ. I grant you that there are in some cases conflicts in the derivation of Λ, but none the less in very different reactions theorists and phenomenologists have been able to derive something in the same range of about 50 MeV to about 400 MeV (Fig. 1), as I showed earlier. In no way, though, do I want to become involved in the way these Λ's are calculated.

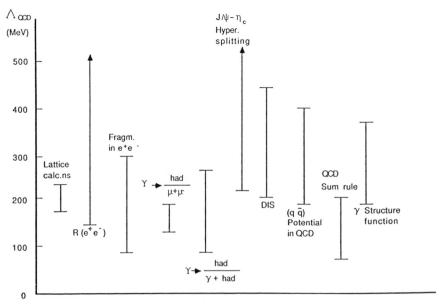

Fig. 1

166

CHEN:

In all the e^+e^- data, and also in the Υ decay, it is always $\Lambda_{\overline{MS}}$ that is being measured.

MISHRA:

You showed us a graph in which $\langle E_{charged}\rangle/\langle E_{total}\rangle \approx 0.5$.
Now, we know from ν–N experiments that

$$3N_f/(16 + N_f) = \langle x_{Bj}^{valence}\rangle/\langle x_{Bj}^{total} = 1\rangle \approx 0.5 \; ,$$

where

$$x = \text{fractional momentum}$$
$$N_f = \text{number of flavours}$$

Should the first result I stated be taken as a consequence of the second? Are they really the same statement? If so, I should consider it an impressive agreement.

ZICHICHI:

Here is the graph I showed (Fig. 2), in which the average energy associated with the charged particles over the total energy is slightly above 0.5. There is an interesting relation that has to do

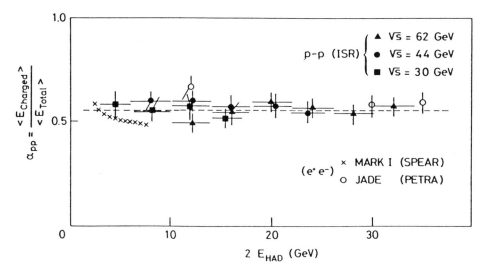

Fig. 2

with your ratio of the average fractional momenta, but it has no connection with this graph. Instead, it has to do with the fact that in ν–p scattering the target proton shows the same leading effect as in pp scattering. The graph we should be looking at is the one in Fig. 3 containing $d\sigma/dx_F$, where you see that the average x_F of the proton is 0.5. The neutrino data show that the valence over the total fractional momentum is 0.5, which I think means that in the deep-inelastic scattering we do have the same leading-particle effect as the one in pp processes. This was a result we proved two years ago: that the leading-particle effect is present in all reactions in the same way—including deep-inelastic scattering. We analysed the leading effect in a reaction in which a neutrino hit a proton, producing a Λ plus anything. We analysed bubble chamber results to see if the Λ was as leading as expected, and we found that this was indeed the case. The reason we used Λ particles was because the Λ was the only baryon that could be identified in the experiment—the authors could not identify protons.

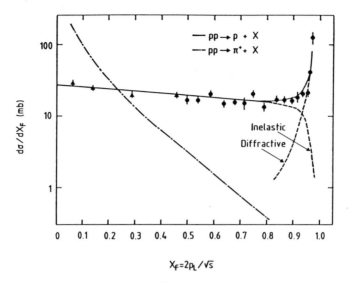

Fig. 3

MARAGE:

The most impressive plot you showed this morning was the prediction you presented of charged-particle multiplicity distributions at the p$\bar{\text{p}}$ Collider. I understand you derived it from ISR dσ/dx, $\langle n_{ch} \rangle$ as a function of effective energy, and one single universal n/$\langle n_{ch} \rangle$ distribution. You would thus not agree with the claim that 'scaling violation effects' are observed at the p$\bar{\text{p}}$ Collider — as was recently made at Bari by the UA5 Collaboration — because, as you pointed out, it appears that the multiplicity distributions in e$^+$e$^-$ and pp ISR experiments are different. Professor Chen attributed this difference to the contribution of heavy flavours in e$^+$e$^-$, and you mentioned the gluon contribution. If it is the latter case, then one should not expect good agreement between the ISR and p$\bar{\text{p}}$ data, since the gluon contributions are different.

ZICHICHI:

You should realize that with the ISR and Collider energies the contribution of heavy quarks is not that much. But the data speak for themselves. We assume that the dσ/dx curve in Fig. 4 is valid at all energies and use these as our first input data.

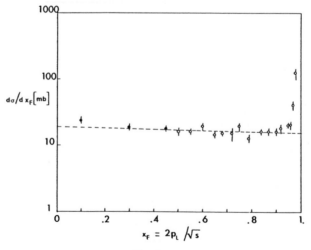

Fig. 4

We take the multiplicity distributions measured at ISR energies (Fig. 5). These show scaling. We assume that this is true at all energies. Finally, for $\langle n_{ch} \rangle$ we take the values obtained using the same extrapolation procedure.

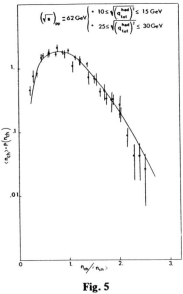

Fig. 5

The data at 14 GeV are shown in Fig. 6. Notice that here the results are expressed in terms of the nominal $\sqrt{s}$. You could ask me why the analysis has not been done using the leading method—it is because we are using other people's data. Figure 7 shows the 30 GeV data where we are using the $x = 0.8$ cut, as is done by the experimenters. For comparison, the dotted line shows our prediction using all x values.

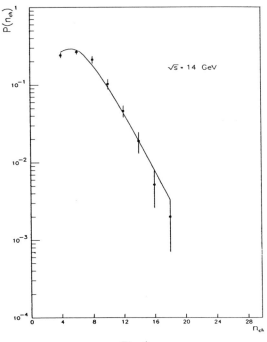

Fig. 6

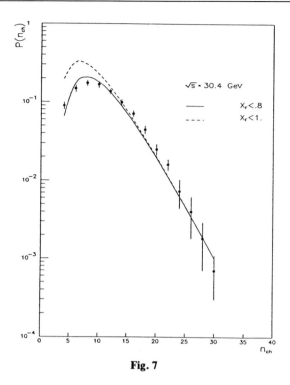

Fig. 7

The 62 GeV and 540 GeV distributions are shown in Figs. 8 and 9, respectively. The agreement between our predictions and the experimental data is spectacular. These results show that the data from Fermilab, the ISR, and UA5 need no scale violation. In fact, using nothing but the measured $d\sigma/dx$ for the proton and the charged-particle distribution at the ISR — assuming

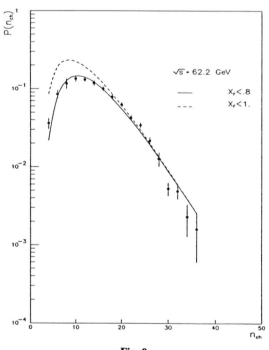

Fig. 8

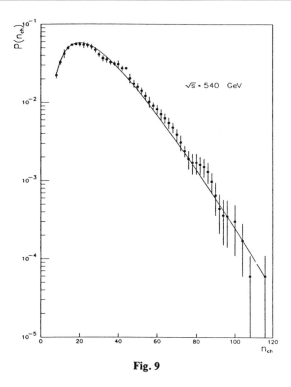

Fig. 9

that they are the same at Collider energy and that $\langle n_{ch} \rangle$ can be extrapolated using our fit—we are able to predict what people observe. So our claim is that the data show the validity of our analysis and that there is no scaling violation.

WOLIN:

I have two questions: the first is, What is the exact experimental definition that you use for leading particles?

ZICHICHI:

I do not have here the data to answer you question in a quantitative way, but it is very simple—it is the fastest particle in the assembly. We did not even have to use Cherenkov counters to identify the protons. Our experiment costs us hardly any money; it was a sort of parasite experiment due to the fact that nature has been very kind in producing a tremendous leading effect—so tremendous that you do not even need to identify the proton. We used a sort of pseudo bubble chamber device—the so-called Split-Field Magnet—to measure the momentum and directions of the particles. This would not give us their identity. However, we know from other results that if you stay in a certain x_F region, the particles ought to be protons and not pions. Look at Fig. 3 again for $d\sigma/dx_F$. Near $x_F = 0.23$, where the two curves intersect, you can see that there are equal numbers of pions and protons. So if you wanted to pick out the leading object in this x_F region without Cherenkov counters, you would have 100% contamination. But if you go out to where $x_F = 0.8$ there is only 2% contamination and you do not need Cherenkov counters. What we wanted to do after all our analysis was to go back and do a powerful experiment with Cherenkov counters and time-of-flight systems to identify the particles and to pick out the real proton in the low-x_F range, which corresponds to high values of the effective energies. But as you know, the ISR has been closed down.

WOLIN:

My second question is, In the two-particle rapidity correlation function you showed earlier, what is the physical interpretation of the variable you plotted and what is the physical significance of the experimental results? I do not mean in the sense that there is agreement between $e^+ e^-$ and hadron–hadron scattering.

ZICHICHI:

The physical interpretation and the significance of the analogy are left to your imagination. The meaning of our result is that when you produce a multihadronic system and study the correlation functions, in spite of the fact that you have totally different initial conditions—such as pp or e^+e^-—the correlation functions show the same behaviour. I want to emphasize that our results do not imply the use of any model. I share your suspicion of this result that the correlation functions look the same, because we know there is something that ought to be different. The relation between the cross-section for heavy-quark production in e^+e^- annihilation and the similar cross-section for light-quark production is a proportionality that just depends on the charges of the quarks. This is not the case, however, in hadronic interactions. Nevertheless, the results show universality features. But the interpretation is left to everyone's imagination. My personal understanding is that if the analysis could be made more precise, there ought to be differences. But to within the errors of our experiment, the reactions are degenerate.

SIMKOVIC:

Professor Zichichi, you have shown that the parameter Λ depends strongly on the values of the W and Z^0 masses and the Weinberg angle. What are the experimental possibilities of improving the accuracy of these parameters? Can we improve, for example, the values of the W mass by measuring radiative corrections, as was originally done in the case of the electron in QED?

ZICHICHI:

I showed you this graph (Fig. 10) this morning. I wanted to show it to you as a beautiful example of how different domains of physics can be related. In spite of the fact that SU(5) is

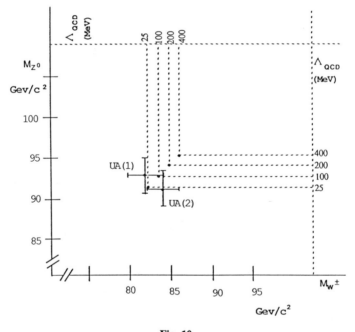

Fig. 10

dead—as Professor Ellis has pointed out—because the proton lifetime is certainly not as predicted, there is still a relation, as implied by this graph, between these particles and Λ, and this fact should be admired. When I was your age the fundamental interactions were like boxes without communication. But I did not want to imply that you should make better measurements of the W and Z^0 to get Λ. I do not think that the value of Λ should be the main purpose of the W and Z^0 analysis. However, if people want to speculate on this, then I would be very happy.

From: QUARKS, LEPTONS, AND THEIR CONSTITUENTS
Edited by Antonino Zichichi
(Plenum Publishing Corporation, 1988)

THE END OF A MYTH:

HIGH-p_T PHYSICS

M. Basile, J. Berbiers, G. Cara Romeo, L. Cifarelli,

A. Contin, G. D'Ali, C. Del Papa, P. Giusti, T. Massam,

R. Nania, F. Palmonari, G. Sartorelli, M. Spinetti,

G. Susinno, L. Votano and A. Zichichi

Bologna-CERN-Frascati Collaboration

Presented by A. Zichichi

1.   INTRODUCTION

So far, the main picture of hadronic physics has been based on a distinction between high-p_T and low-p_T phenomena.

In the framework of parton model, high-p_T processes were the only candidates to establish a link between

- purely hadronic processes
- (e^+e^-) annihilations
- (DIS) processes .

The advent of QCD has emphasized in a dramatic way the privileged role of high-p_T physics due to the fact that, thanks to asymptotic freedom, QCD calculations via perturbative methods can be attempted at high-p_T and results successfully compared with experimental data [1]. The conclusion was: we can forget about everything else and limit ourselves to high-p_T physics.

Being theoretically off limits, low-p_T phenomena, which represent the overwhelming majority of hadronic processes (more than 99%

2   A. ZICHICHI ET AL.

of physics it is here), have been up to now neglected. By subtracting
the leading proton effects in order to derive the effective energy
available for particle production and by using the correct variables,
the BCF collaboration has performed a systematic study of the final
states produced in low-p_T (pp) interactions at the ISR and has compared the results with those obtained in the processes listed below:

Process	Data Sources
(e^+e^-)	SLAC, DORIS, PETRA
(DIS)	SPS/EMC
(pp)	ISR (AFS)
$(\bar{p}p)$	SPS Collider (UA1)
(e^+e^-)	PETRA/TASSO (leading subtraction)

where (pp) and $(\bar{p}p)$ are bracketed under "Transverse physics", and ISR (AFS) and SPS Collider (UA1) are bracketed together.

The results of this study [2-18] show that, once a common basis
for comparison is found by the use of the correct variables, remarkable analogies are observed in processes so far considered basically
different like

- low-p_T (pp) interactions
- (e^+e^-) annihilations
- (DIS) processes
- high-p_T (pp) and $(\bar{p}p)$ interactions

This is how universality features emerge, and this is the basis
to proceed for a meaningful comparison, i.e.:

first identify the correct variables to establish a common basis,
then proceed to a detailed comparison[*].

1.1 The leading effect

I would like to show you the importance of the leading proton
effect at the ISR.

---

[*] The root of this new approach to the study of hadronic interactions goes back a long time to a proposal by the CERN-Bologna
group: "Study of deep inelastic high momentum transfer hadronic
collisions" PMI/com-69/35, 8 July 1969.

In Figure 1 is reported the $d\sigma/dx_F$ measured in proton-proton (pp) interactions (x_F indicates the usual Feyman x); the data scale at ISR energies [$(\sqrt{s})_{pp}$ = 23-62 GeV] and are integrated over all p_t values.

This distribution can be divided into three different main parts: the diffractive and elastic region ($x_F > 0.8$); the central region ($x_F < 0.2$); the intermediate region ($0.2 < x_F < 0.8$). This last region has been until now considered uninteresting and consequently neglected. On the contrary, all I am going to say from now on is exactly located in this particular interval.

Notice the dramatic difference between the way by which protons and pions are produced.

$$\frac{d\sigma}{dx_F} \approx \text{const. for p (\underline{leading})}$$

and

$$\frac{d\sigma}{dx_F} \propto (1 - x)^3 \text{ for } \pi \text{ (\underline{not leading})}$$

Let me stress that the leading effect is not limited to the ISR case, nor to (pp) collisions.

We have investigated this phenomenon and we have found that the leading hadron effect is there whenever a hadron is present in the initial state, no matter if the interaction is initiated by a hadron, or by a photon, or by a weak boson [2,3].

To make a quantitative estimate of the leading hadron effect we have introduced the quantity

$$L(x_0, x_1, x_2)$$

defined as

$$L(x_0, x_1, x_2) = \frac{\int_{x_1}^{x_2} F(x) \, dx}{\int_{x_0}^{x_1} F(x) \, dx}$$

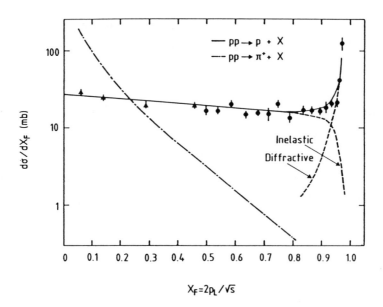

Fig. 1   pp interactions: longitudinal momentum distribution for proton (full line) and pions (dotted line)

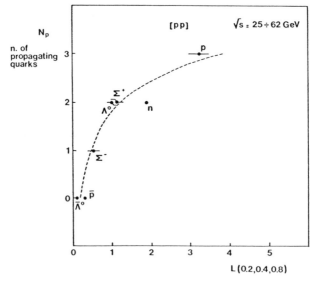

Fig. 2   The 'leading' quantity L(0.2, 0.4, 0.8) dirived for different types of baryons produced in (pp) collisiosn at the CERN ISR. The hadrons are ordered according to the number of propagating quarks. The dashed curve superimposed is obtained by using a parametrization of the single-particle inclusive cross-section, $F(x) = (1 - x)^{2n-1}$ where n is the number of quarks changed by the incident hadron.

with

$$F(x) = \frac{1}{\pi} \int \frac{2E}{\sqrt{s}} \frac{d\sigma}{dx \, dp_T^2} \, dp_T^2$$

where $F(x)$ is the inclusive single particle invariant cross-section integrated over p_T.

In order to exclude central production ($x_F < 0.2$) and the dif-fractive processes ($x_F > 0.8$), we have chosen

$$x_0 = 0.2$$
$$x_1 = 0.4$$
$$x_2 = 0.8 \; .$$

Please note that these limits have been singled out for conve-nience, their choice is not particularly relevant for the future considerations.

As a result of this analysis, the leading effect has been found to be more pronounced when all quarks are allowed to go from the initial to the final state. If 2 out of 3 quarks are allowed to go, the effect diminishes. If 1 out of 3 quarks is allowed to go, it is still less.

This is shown, for the ISR case, in Fig. 2, where the quantity $L(x_0, x_1, x_2)$ for various particles in the final state is plotted versus the number of propagating quarks. As a consequence of the above mentioned choice of the limits, the quantity L is equal to 3 when 3 quarks propagate (as in the case of the proton); to 1.5 when only 2 quarks propagate (Λ^0, Σ^+, n); to 0.3 when no quarks are propagating.

Going on in the search for a leading hadron effect, we have analysed other groups results on the Λ^0 productions in (ep) and ($\bar{\nu}p$) interactions (Fig. 3). We have found that the quantity L is the same.

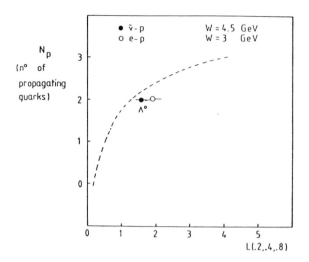

Fig. 3 L(0.2, 0.4, 0.8) for Λ^0 productions in (v̄p) and (ep) reactions. The dashed line is the same as that described for fig. 2.

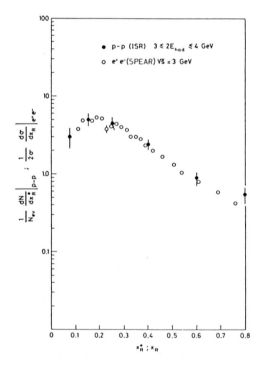

Fig. 4 The inclusive single-particle fractional momentum distributions $(1/N_{ev}) \cdot (dN/dx_R^*)$ in the interval 3 GeV $\leq 2E_{had} \leq$ 4 GeV obtained from data at $\sqrt{s}$ = 30 GeV. Also shown are data from MARK I at SPEAR.

1.2 <u>Kinematic diagrams and identification of the correct variables</u>

Now it is important to identify the correct variables in describing hadron production in (pp) interactions, (e^+e^-) annihilations and DIS processes. This will be the basic starting point to put on <u>equal footing</u> these three ways of producing multiparticle hadronic systems [4].

1.2.1 <u>No leading subtraction: (e^+e^-)</u>

With reference to the following diagram

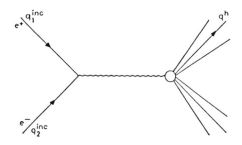

we define the quantities:

$$(q_1^{inc} + q_2^{inc}) \equiv (q_{inc}^{inc})_{tot}$$

$$(x_{inc}^{inc}) = \frac{q_1^h \cdot (q_{inc}^{inc})_{tot}}{(q_{inc}^{inc})_{tot}^2} \ .$$

If $\vec{p}_1^{\,inc} = -\vec{p}_2^{\,inc}$ these quantities are equal to:

$$(q_{inc}^{inc})_{tot}^2 = (q_1^{inc} + q_2^{inc})^2 = 4(E^{inc})^2$$

$$(x_{inc}^{inc}) = \frac{q^h \cdot (q_1^{inc} + q_2^{inc})}{(q_1^{inc} + q_2^{inc})^2} \simeq \frac{2E^h \cdot E^{inc}}{4(E^{inc})^2} = \frac{E^h}{2E^{inc}} \ .$$

Since the minimum number of final states particles is 2, we can write

$$(x_{inc}^{inc}) \doteq \frac{E^h}{E^{inc}} .$$

1.2.2 One leading subtraction: (DIS)
With reference to the following diagram

$$q_1^{had} \equiv (i p_1^{had}; \nu \equiv E_1^{had}) .$$

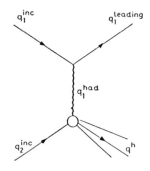

we can here define the quantities:

$$(q_1^{had} + q_2^{inc}) \equiv (q_{inc}^{had})_{tot}$$

$$(x_{inc}^{had}) = \frac{q^h \cdot (q_{inc}^{had})_{tot}}{(q_{inc}^{had})_{tot}^2} .$$

with few approximations we obtain

$$(q_{inc}^{had})_{tot}^2 = (q_1^{had} + q_2^{inc})^2 \doteq 2E_1^{had} \cdot E_2^{inc}$$

$$(x_{inc}^{had}) = \frac{q^h \cdot (q_1^{had} + q_2^{inc})}{(q_1^{had} + q_2^{inc})^2} \doteq \frac{E^h \cdot E_2^{inc}}{2E_1^{had} \cdot E_2^{inc}} = \frac{E^h}{2E_1^{had}}$$

even in this case, as the minimum number of final state particles is 2, we otbain

$$(x_{inc}^{had}) \doteq \frac{E^h}{E^{had}} \ .$$

Note that the standard DIS analysis make use of the quantities

$$W^2 = (q_1^{had} + q_2^{inc})^2 \doteq 2E_1^{had} \cdot E_2^{inc} \ .$$

$$Z = \frac{q^h \cdot q_2^{inc}}{q_1^{had} \cdot q_2^{inc}} \doteq \frac{E^h \cdot E_2^{inc}}{E_1^{had} \cdot E_2^{inc}} = \frac{E^h}{E_1^{had}}$$

$$Q^2 = (q_1^{had})^2 \ .$$

1.2.3 Two leadings: (pp)

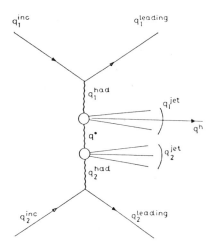

From the above diagram we can here define the quantities:

$$(q_1^{had} + q_2^{had}) \equiv (q_{had}^{had})_{tot}$$

$$(x_{had}^{had}) = \frac{q^h \cdot (q_{had}^{had})_{tot}}{(q_{had}^{had})_{tot}^2} \; .$$

With standard approximations and in the case $E_1^{had} = E_2^{had}$, we can write

$$(q_{had}^{had})_{tot}^2 = (q_1^{had} + q_2^{had})^2 \doteq 4E_1^{had} \cdot E_2^{had} \approx 4(E^{had})^2$$

$$(x_{had}^{had}) = \frac{q^h \cdot (q_1^{had} + q_2^{had})}{(q_1^{had} + q_2^{had})^2} \doteq \frac{E^h \cdot (E_1^{had} + E_2^{had})}{4E_1^{had} \cdot E_2^{had}} \approx \frac{E^h}{2E_1^{had}}$$

since we have at least 2 particles in the final state, we have

$$(x_{had}^{had}) \approx \frac{E_1^h}{E^{had}} \; .$$

1.2.4 Two bodies interactions

With reference to the diagrams shown in the previous paragraphs, let me remind you, for each interaction, the definition of the variables conventionally used in the study of these processes.

- (pp), (p̄p), (πp), (Kp), etc.

  no leading subtraction

$$x = \frac{E^h}{E_{1,2}^{inc}} \; ; \qquad (\sqrt{s}) \; \begin{array}{l} pp \\ \pi p \\ Kp \\ \cdot \\ \cdot \\ \cdot \end{array} \equiv 2 \sqrt{(E_1^{inc} \cdot E_2^{inc})} \; .$$

- (μp), (ep), (γp), etc.

  one leading subtraction

  $$Z = \frac{E^h}{E_1^{had}} \; ; \qquad (\sqrt{W^2}) \; \begin{matrix} \mu p \\ ep \\ \gamma p \\ \cdot \\ \cdot \\ \cdot \end{matrix} \equiv \sqrt{(2E_1^{had} \cdot E_2^{inc})} \; .$$

- $(e^+ e^-)$ annihilation

  two leading subtraction

  $$x = \frac{E^h}{E_1^{had}} \; ; \qquad (\sqrt{s})_{e^+ e^-} \equiv 2\sqrt{(E_1^{had} \cdot E_2^{had})} \; .$$

Analysis based on the use of these variables failed to establish connections among these processes. The reason for this failure
lies in the fact that the leading particles (when present) have not
been subtracted [5]; only with this step it is possible a comparison
among these apparently different reactions.

In fact, not all the energy carried by the incident proton can
be used to produce particles, since we know that about 50% of it is
dragged out by the outgoing leading proton. So the actual energy
available for particle production is

$$\sqrt{(q_{tot}^{had})^2} = (q_1^{had} + q_2^{had})$$

where

$$q_i^{had} = q_i^{inc} - q_i^{lead} \; .$$

$\sqrt{(q_{tot}^{had})^2}$ is now the correct variable to compare with $(\sqrt{s})_{e^+ e^-}$ [*].

---

[*] In the following, some data will be presented plotted as a function of $2E^{had} = 2(E^{inc} - E^{leading})$. This quantity, used at an
   early stage of the analysis, is essentially equivalent to
   $\sqrt{(q_{tot}^{had})^2}$

12   A. ZICHICHI ET AL.

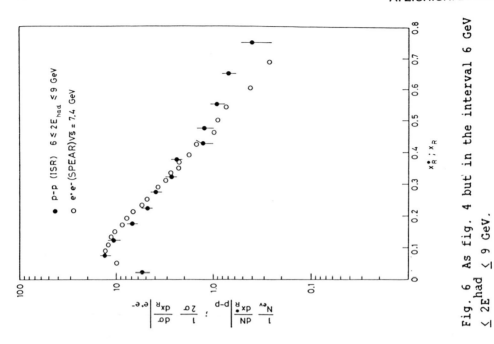

Fig. 6 As fig. 4 but in the interval 6 GeV $\leq 2E_{had} \leq 9$ GeV.

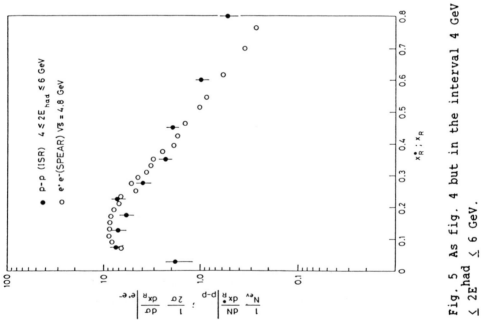

Fig. 5 As fig. 4 but in the interval 4 GeV $\leq 2E_{had} \leq 6$ GeV.

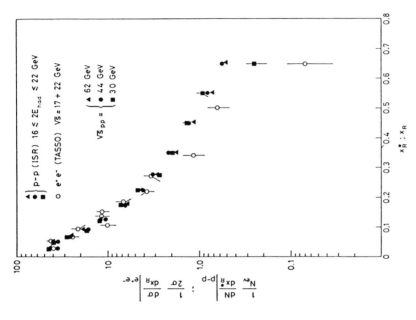

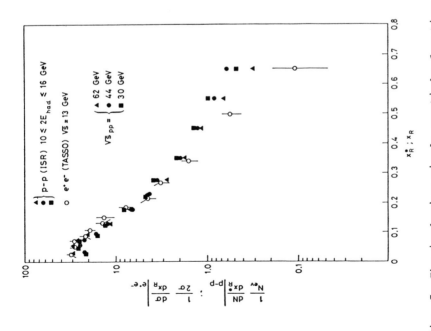

Fig. 7 The inclusive single-particle fractional momentum distributions $(1/N_{ev}) \cdot (dN/dx_R^*)$ in the interval 10 GeV $\leq 2E_{had} \leq$ 16 GeV obtained from data at different $(\sqrt{s})_{pp}$. Also shown are data from TASSO and PETRA.

Fig. 8 As fig. 7 but in the interval 16 GeV $\leq 2E_{had} \leq$ 22 GeV.

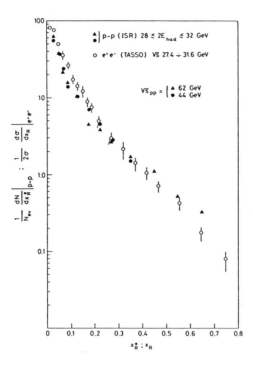

Fig. 9   As fig. 7 but in the interval 28 GeV $\leq$ 2E^{had} $\leq$ 32 GeV.

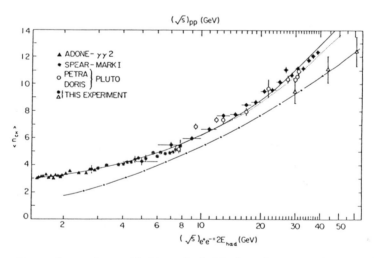

Fig. 10 Mean charged-particle multiplicity [averaged over different $(\sqrt{s})_{pp}$] vs. 2E^{had}, compared with (e$^+$e$^-$) data. The continuous line is the best fit to our data according to the formula $\langle n_{ch} \rangle$ = a + b exp $\{c\sqrt{[\ln (s/\Lambda)^2]}\}$. The dotted line is the best fit using PLUTO data. The dash-dotted line is the standard (pp) total charged particle multiplicity with, superimposed, our data as open triangular points.

I would like to call your attention on two facts. Firstly, that the standard quantity $(\surd s)_{pp}$ is just a superposition of various $\surd(q_{tot}^{had})^2$; secondly, that the only thing we need to calculate $\surd(q_{tot}^{had})^2$ are the two outgoing leading protons.

To compare our data with DIS, we will instead make use of the quantity

$$(W^2)_{pp}^{had} = (q_1^{had} + q_2^{inc})_{pp}^{2} \equiv (W^2)_{DIS} \; .$$

Please note that in the latter case we will subtract only one leading proton.

2.   EXPERIMENTAL RESULTS: COMPARISON WITH e$^+$e$^-$

The data were selected from 10^5 interactions collected at the Split Field Magnet (SFM) spectrometer at the ISR.

Three different energies were scanned: 30, 44, 62 GeV; the leading proton was experimentally identified as the fastest particle in the hemisphere with positive charge and $0.4 < x_F < 0.8$.

In the following paragraphs, I will use the leading subtraction method to compare various properties which characterise the multi-particle systems produced in (pp) collisions with (e$^+$e$^-$).

2.1   dσ/dx$_F$

Figures 4 to 9 show the fractional longitudinal energy distribution in (e$^+$e$^-$) and in proton-proton interaction, at $(\surd s)_{e^+e^-} = \surd(q_{tot}^{had})^2$ for different energy intervals. Note that different ISR centre of mass energy (i.e. 30, 40, 62 GeV) gives the same results once we have fixed $\surd(q_{tot}^{had})^2$. The remarkable agreement proves the validity of our new approach* [5-8].

---

* The little discrepancies present in the tail of the distribution, especially for high $\surd(q_{tot}^{had})^2$ can be accounted for the pions contamination we have in the leading particle definition.

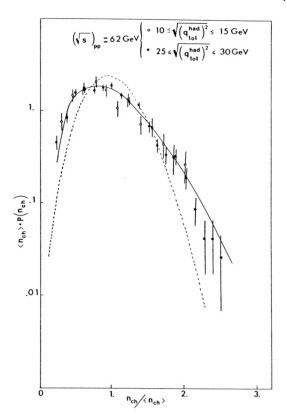

Fig. 11 Multiplicity distributions as measured in pp collisions at two $\int(q_{tot}^{had})^2$ values compared with (e^+e^-) data (dashed line).

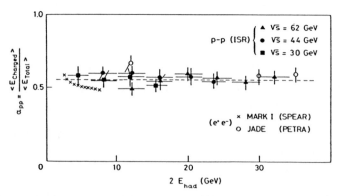

Fig. 12 The charged-to-total energy ratio α_{pp}, obtained in (pp) collisions plotted vs. $2E^{had}$ and compared with (e^+e^-) obtained at SPEAR and PETRA.

2.2 $\langle n_{ch} \rangle$

Figure 10 shows the charged particles multiplicity measured in
(pp), (e^+e^-) and (pp) with the leading protons subtracted. The
agreement in the last two types of data is complete [9-11].

2.3 Multiplicity distributions

Figure 11 shows the multiplicity distributions measured for two
different $\int (q_{tot}^{had})^2$ and compared with (e^+e^-). Notice that here we
find the first difference: (e^+e^-) shows a narrower distribution with
respect to $\int (q_{tot}^{had})^2$. In any case the pp data scale with $\int (q_{tot}^{had})^2$ [12].
This observation will be used subsequently in our study of the
leading effect of the Sp$\bar{p}$S.

2.4 $\langle E_{ch} \rangle / E_{total}$

Figure 12 shows the ratio of the average energy carried by the
charged particles to the total energy available, as measured in pp
interactions as a function of $\int (q_{tot}^{had})^2$. The (e^+e^-) data are also
reported, and agree with our measurements [13].

2.5 p_t^2 distributions

From now on, I will use the notation p_T to indicate the transverse momentum relative to the beam axis, and p_t to indicate the
transverse momentum relative to the jet axis. In our procedure the
jet axis is directly deduced from the leading protons kinematical
quantities.

In Figures 13 to 18 we report the $(p_t)^2$ distributions at
various $\int (q_{tot}^{had})^2$ compared with equivalent $(\sqrt{s})_{e^+e^-}$.

We find that at high $\int (q_{tot}^{had})^2$ the tails of the (e^+e^-) distributions are somehow higher than our data. This phenomenon is a
reflection of the fact that $\langle p_t^2 \rangle$ does not increase with $\int (q_{tot}^{had})^2$,
while it does increase with $(\sqrt{s})_{e^+e^-}$ [14].

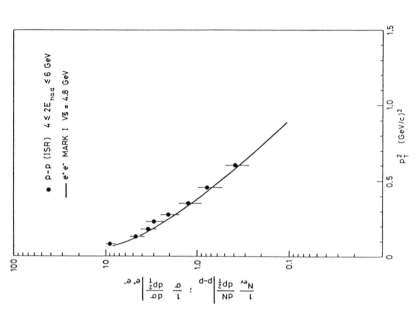

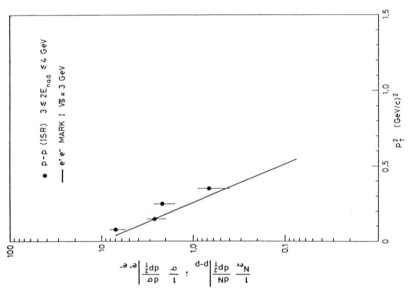

Fig. 14 As fig. 13 but in the interval 4 GeV $\leq$ 2E$_{had}$ $\leq$ 6 GeV.

Fig. 13 The inclusive single-particle transverse momentum distribution $(1/N_{ev}) \cdot (dN/dp_T^2)$ in
the interval 3 GeV $\leq$ 2E$_{had}$ $\leq$ 4 GeV obtained from
data at $\sqrt{s}$ = 30 GeV. Also shown in the fit to
the SPEAR data (continuous line).

END OF A MYTH: HIGH-p_T PHYSICS   19

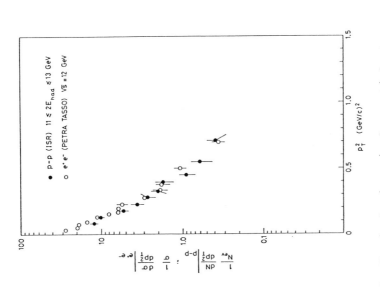

Fig. 16 The inclusive single-particle transverse-momentum distributions $(1/N) \cdot (dN/dp_T^2)$ in the interval 11 GeV $\leq 2E_{had} \leq$ 13 GeV obtained from data at $\sqrt{s}$ = 62 GeV. Also shown are data from TASSO at PETRA.

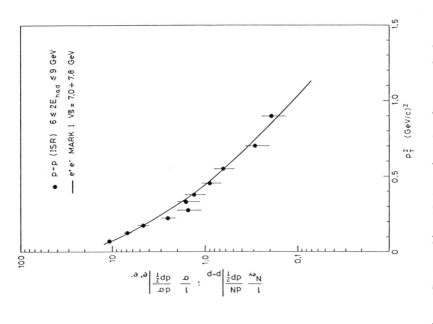

Fig. 15 As fig 13 but in the interval 6 GeV $\leq 2E_{had} \leq$ 9 GeV.

20   A. ZICHICHI ET AL.

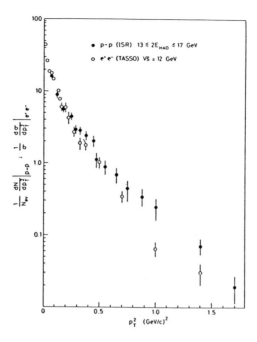

Fig. 17   As Fig. 16 but in the range 13 GeV $\leq$ 2E^{had} $\leq$ 17 GeV.

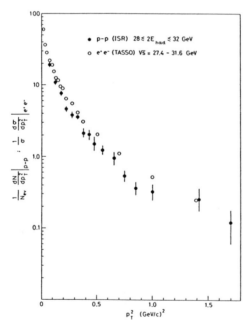

Fig. 18   As Fig. 16 but in the range 28 GeV $\leq$ 2E^{had} $\leq$ 32 GeV.

2.6 $p_t/\langle p_t \rangle$

Although the rise of p_t^2 with the energy is less pronounced in the (pp) case than in (e^+e^-) case, the agreement is good, even at the highest energy, if the variable $p_t/\langle p_t \rangle$ is used, where $\langle p_t \rangle$ is the average p_t of all produced particles.

The distribution of $p_t/\langle p_t \rangle$ is expected from QCD to scale with energy, and this is indeed observed in (e^+e^-) data as well as in our (pp) data, as shown in Fig. 19 [15].

2.7 The event planarity: $\langle p_t^2 \rangle_{in,out}$

The planarity of the events has been studied by means of the $\langle p_t^2 \rangle_{in,out}$-distributions.

These quantities are determined in the following ways.

a) Construct the tensor

$$M_{\alpha\beta} = \sum_{j=1}^{N} P_{j\alpha} \cdot P_{j\beta} \quad (\alpha,\beta = 1,2)$$

where N ≡ number of particles in the event.

b) Determine the eigenvectors $\vec{n}_1$, $\vec{n}_2$ and the eigenvalues Λ_1, Λ_2 ($\Lambda_1 < \Lambda_2$).

c) Define

$$\langle p_t^2 \rangle_{out} = \frac{\Lambda_1}{N} = \sum_{j=1}^{N} (\vec{p}_j \cdot \vec{n}_1)^2$$

$$\langle p_t^2 \rangle_{in} = \frac{\Lambda_2}{N} = \sum_{j=1}^{N} (\vec{p}_j \cdot \vec{n}_2)^2 .$$

The results of this analysis are shown in Figs. 20 and 21 together with a comparison with (e^+e^-). Note that at the highest energy (e^+e^-) show a higher tail in the $\langle p_t^2 \rangle_{in}$ as well as in the $\langle p_t^2 \rangle_{out}$ [16].

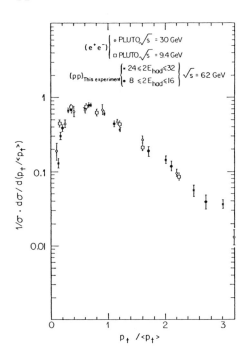

Fig. 19 The differential cross-section $(1/\sigma)[d\sigma/d(p_t/\langle p_t \rangle)]$ vs. the 'reduced' variable $p_t/\langle p_t \rangle$. These distributions allow a comparison of the multiparticle systems produced in (e^+e^-) annihilation and in (pp) interactions in terms of the renormalized transver transverse-momentum properties.

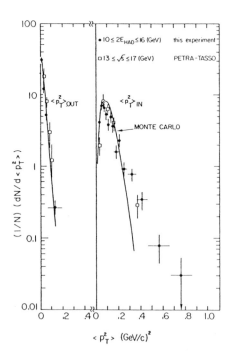

Fig. 20 The average transverse-momentum squared distributions for 'out' and 'in' cases, in the 'low-energy' range.

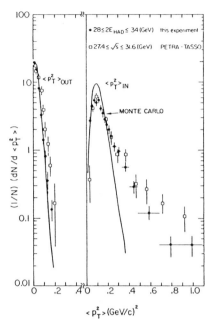

Fig. 21 The average transverse-momentum squared distributions for 'out' and 'in' cases, in the 'high-energy' range.

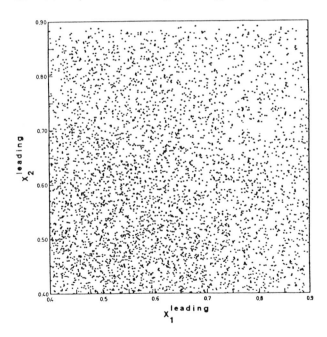

Fig. 22 Scatter plot of the fractional energies of the two leading protons, $x_{1,2}^{leading} = E_{1,2}^{leading}/E^{inc}$, in the range of $0.4 \leq x_{1,2}^{leading} < 0.9$, at $(\sqrt{s})_{pp} = 62$ GeV.

2.8 Correlation between leading 1 - leading 2

So far the correlations in (pp) reactions have been studied without subtracting the leading protons.

The results obtained in this way show a difference with the (e^+e^-) case.

Our point is that, as for the other properties, the correlations must be studied once the leading protons are subtracted.

In order to try to understand the correlations in (pp) reactions and the comparison with (e^+e^-) annihilations, the first step is to study the correlation between the two leading protons.

Figure 22 shows the results of this study and proves that the two protons are uncorrelated [17].

A more quantitative analysis of the same data is reported in Fig. 23, where in the abscissa are reported the values of $x_1^{leading}$, and in the ordinate the quantity:

$$\left[\langle x_2^{leading} \rangle \right]_{x_1^{leading}\,fixed} = \left[\frac{\sum\limits_{i=1}^{N} x_2^{leading}}{N} \right]_{x_1^{leading}\,fixed}$$

2.9 Two particle correlations

The two particle correlation function in rapidity space $R(y,y')$ is defined as

$$R(y,y') = \frac{\varrho_2(y,y')}{f\,\varrho_1(y)\,\varrho_1(y')} - 1$$

where

- $\varrho_1(y)$ is the normalized single-particle inclusive rapidity distribution:

$$\varrho_1(y) = \frac{1}{\sigma_{in}} \frac{d\sigma(y)}{dy}$$

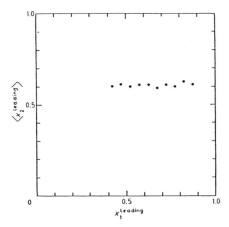

Fig. 23 The average value $\langle x_2^{leading} \rangle$ for fixed $x_1^{leading}$ plotted vs. $x_1^{leading}$.

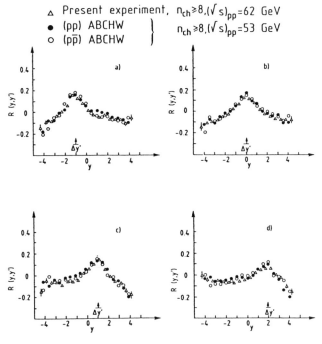

Fig. 24 Two-particle correlations in rapidity space for different y' intervals [a) -1.75 $\leq$ y' $\leq$ -0.75, b) -0.25 $\leq$ y' $\leq$ 0.25, c) 0.75 $\leq$ y' $\leq$ 1.25, d) 1.75 $\leq$ y' $\leq$ 2.25] as measured by the ACBHW collaboration at $(\sqrt{s})_{pp}$ = 53 GeV in (pp) (black points) and ($p\bar{p}$) (open circles) collisions and by the present experiment (open triangles) at $(\sqrt{s})_{pp}$ = 62 GeV in (pp) collisions. These results are obtained from minimum-bias events without leading-proton subtraction for an observed charged multiplicity: $n_{ch} \geq 8$.

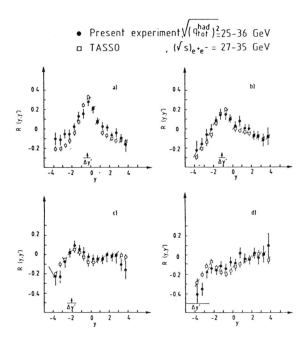

Fig. 25 Two-particle correlation in rapidity space: R(y,y'), for
different y' intervals, as measured in the present experiment after
the leading-proton subtraction in the $\sqrt{(q_{tot}^{had})^2}$ range 25 to 36 GeV
(black points), compared with the results by the TASSO collaboration
at $(\sqrt{s})_{e^+e^-}$ between 27 and 35 GeV (open squares).

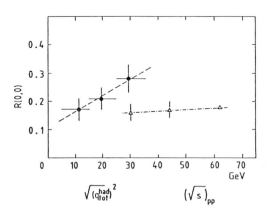

Fig. 26 Correlation function R(y,y') measured at y = y' = 0, vs.
$\sqrt{(q_{tot}^{had})^2}$. The data are indicated as black points. The broken line is
the best fit. For comparison, the open triangles show R(0,0) vs.
$(\sqrt{s})_{pp}$. In this case the analysis of the final state is made without
subtracting the leading proton effects. The dash-dotted line is the
best fit to these data. Notice that the abscissa for the black
points is $\sqrt{(q_{tot}^{had})^2}$; for the open triangles, it is $(\sqrt{s})_{pp}$.

(σ_{in} = inelastic cross-section).

- $\varrho_2(y,y')$ is the two-particle normalized rapidity distribution:

$$\varrho_2(y,y') = \frac{1}{\sigma_{in}} \frac{d\sigma(y,y')}{dydy'}$$

- f is a normalization factor

$$f = \frac{\langle n_{ch} (n_{ch} - 1)\rangle}{\langle n_{ch}\rangle^2} .$$

In the case of no correlation we have

$$R(y,y') = 0 .$$

The measured values of $R(y,y')$ are reported for two different cases:

a) Without subtracting the leading proton 1 (Fig. 24).

b) With our analysis, i.e. subtracting the leading protons (Fig. 25).

The results [18,19] show that:

- in case 1 there is perfect agreement with previous results in (pp) reactions;

- in case 2 there is excellent agreement with (e^+e^-) results.

Figure 26 shows the dependence of $R(0,0)$ on the (pp) c.m. energy $(\sqrt{s})_{pp}$ and on the "effective" hadronic energy $\sqrt{(q_{tot}^{had})^2}$.

Unfortunately, no (e^+e^-) data are available for comparison in this case.

2.10 Scale breaking effects

Scale breaking effects are not expected in simple quark-parton models.

They are predicted to exist by QCD as a consequence of the emission of gluons from the primary quarks.

These effects have been looked for in the multiparticle hadronic states produced in (e^+e^-) annihilation.

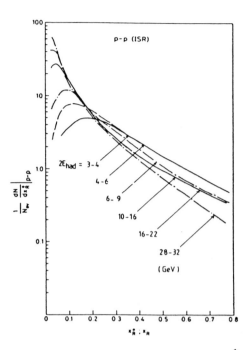

Fig. 27 Fits to the distribution $(1/N_{ev}) \cdot (dN/dx_R^*)$ as measured at the ISR (pp collisions) for different values of $2E_{had}$.

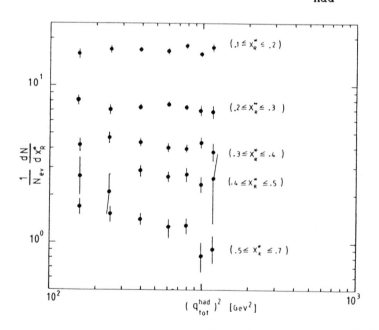

Fig. 28 The quantity $(1/N_{ev}) \cdot (dN/dx_R^*)$ is plotted vs. $(q_{tot}^{had})^2$, for different ranges of x_R^*.

Even in pp interactions we find such an effect as shown in Fig. 27 [20].

A more quantitative way for comparing our results with (e^+e^-) annihilation data, is to fit the data of each x_R^* interval with the function

$$\frac{1}{N_{ev}} \frac{dN}{dx_R^*} = A \left[1 + C \ln (q_{tot}^{had})^2 \right] .$$

(see Fig. 28).

The values of the parameter C vs x_R^* are reported in Fig. 29 together with the values of the same parameter obtained in (e^+e^-) events at PETRA.

Within experimental errors, the values of the parameter C measured in (pp) and (e^+e^-) show the same trend.

This proves that also in the scale breaking effects the multi-particle systems produced in (pp) interactions show striking analogies with the multiparticle systems produced in (e^+e^-) annihilations.

3.   EXPERIMENTAL RESULTS: COMPARISON WITH e^+e^-
     WHEN LEADING EFFECT IS IN

Now I will show what happens in (e^+e^-) when you have a leading particle. The TASSO Group has in fact observed a leading behaviour for the D^* produced in (e^+e^-).

It was the first time that such an observation of a leading effect was made in (e^+e^-).

The simple comparison of the fractional energy distribution of standard (e^+e^-) jet with these events containing D^* shows a dramatic difference (Fig. 30).

On the other hand, once you have subtracted the D^* and compared the remaining jet with a standard e^+e^- jet at equivalent energy, you find a complete agreement (Fig. 31). Note that this agreement extends to our data [(pp) collisions] at equivalent $\sqrt{(q_{tot}^{had})^2}$. The comparison in the variables P_t^2 it is shown in Fig. 32.

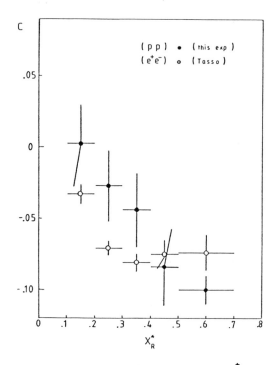

Fig. 29 The values of the coefficient C vs. x_R^*, as measured in (pp) interactions (this experiment) and in (e^+e^-) annihilation.

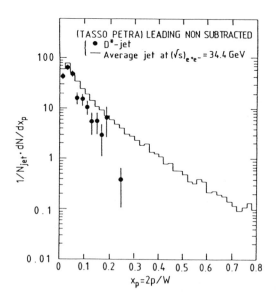

Fig. 30 PETRA-TASSO collaboration. The distribution $(1/N_{jet}) \cdot (dN/dx_p)$ measured in standard jets and in jets containing a D^*.

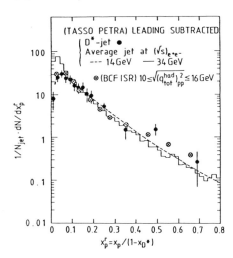

Fig. 31 As fig. 30 but, for the jets containing a D*, the leading charm effect has been subtracted. Also shown are the data from our analysis at the ISR.

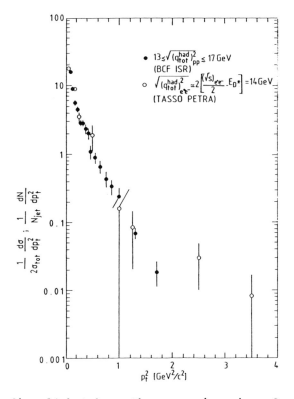

Fig. 32 As fig. 31 but here the comparison is made in the variable p_t^2.

4.   EXPERIMENTAL RESULTS: COMPARISON WITH DIS

In order to compare (pp) data with DIS data we have applied to (pp) data a DIS-like analysis (see paragraph 1.2.4).

The following quantities, which characterise the hadronic system produced, have been measured and compared with DIS data:

a) $\langle n_{ch} \rangle$

b) $d\sigma/dz$.

The analysis will use essentially data at $\sqrt{s}_{pp}$ = 30 GeV, this in order to have a range of W^2 to allow a comparison with DIS results.

4.1   The average charged particle multiplicity
       in (pp) and in (νp) reactions

The data are reported in Fig. 33, where they are compared with (νp) data [21].

In the overlap energy region the agreement is good.

Furthermore, our data at higher energies lie on the extrapolation of the fit to (νp) data.

Figure 34 shows that the agreement with DIS is not only in the total charged multiplicity, but even in the forward and backward regions taken separately [22].

4.2   Comparison of dσ/dz in (pp) and (μp) interactions

We have compared the $(1/N_{ev})$ (dN/dz) distributions in (pp) and (μp) reactions at two $(W^2)_{\mu p}$ values.

The results, shown in Figs. 35 and 36, exhibit again good agreement of (pp) and (μp) data [23].

5.   FIRST CONCLUSION ON (e^+e^-) AND DIS

All the results presented have been obtained in low-p_T (pp) interactions, using

$$\sqrt{(q^{had}_{tot})^2}$$

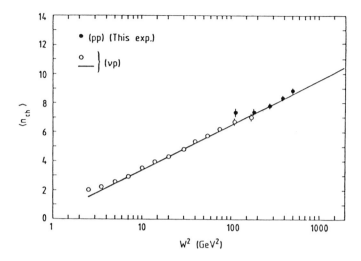

Fig. 33 The average charged-particle multiplicities $\langle n_{ch} \rangle$ measured in (pp) at $(\sqrt{s})_{pp}$ = 30 GeV, using a DIS-like analysis, are plotted vs. W^2 (black points). The open points are the (vp) data and the continuous line is their best fit.

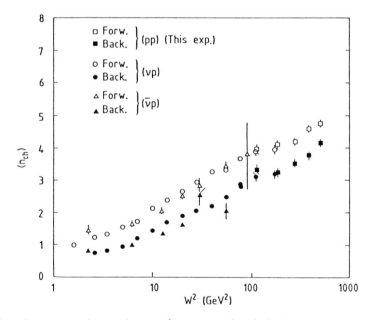

Fig. 34 The mean charged-particle multiplicities $\langle n_{ch} \rangle_{F,B}$ in the forward and backward hemispheres vs. W^2, in (vp), $(\bar{v}p)$ and (pp) interactions.

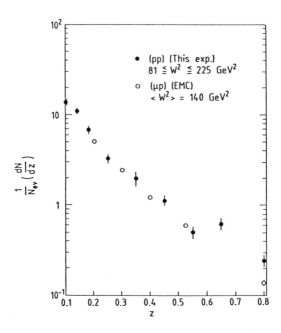

Fig. 35 The inclusive distribution of the fractional energy z for (pp) reactions in the energy interval 81 (GeV)^2 $\leq$ W^2 $\leq$ 225 (GeV)^2 compared with the data from (μp) reactions at ⟨W^2⟩ = 140 (GeV)^2.

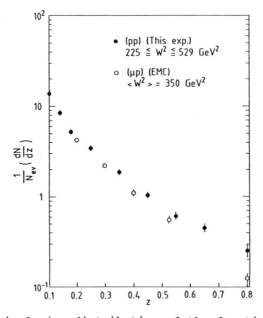

Fig. 36 The inclusive distribution of the fractional energy z for (pp) reactions in the energy interval 225 (GeV)^2 $\leq$ W^2 $\leq$ 529 (GeV)^2 compared with the data from (μp) reactions at ⟨W^2⟩ = 350 (GeV)^2.

for the comparison with (e$^+$e$^-$) and

$$\sqrt{(q_1^{had} + q_2^{inc})^2} .$$

for the comparison with DIS.

These results show that it <u>is not true that only high-p$_T$ reac-</u><u>tions can be compared with (e$^+$e$^-$) and DIS processes and this is the</u> <u>end of a myth.</u>

We can now turn ourselves to compare our low-p$_T$ physics with the high-p$_T$ physics at ISR and Sp$\bar{\text{p}}$S.

6.   EXPERIMENTAL RESULTS:

COMPARISON WITH TRANSVERSE PHYSICS AT ISR

In (pp) interactions the same values of

$$\sqrt{(q_{tot}^{had})^2}$$

can be produced in two ways [24]:

 i) At low-p$_T$: we have seen the striking analogies between the pro-
    perties of the multihadronic final states produced in (pp),
    (e$^+$e$^-$) and DIS (striking analogies apart from $\langle p_t \rangle$ and multi-
    plicity distribution).

ii) At high-p$_T$: we have not been able to study, at constant values of
    $\sqrt{(q_{tot}^{had})^2}$, the multihadronic systems produced at high p$_T$, the
    main reason being lack of ISR machine time. Therefore a compa-
    rison between multihadronic final states at high-p$_T$ and low-p$_T$
    in (pp) interactions at ISR energies has been impossible.

It is possible that, had we been allowed to study high-p$_T$ multihadronic systems at the ISR, we would have found the value of $\langle p_t^2 \rangle$
to be identical to the value measured in (e$^+$e$^-$) for

$$(\sqrt{s})_{e^+ e^-} = \sqrt{(q_{tot}^{had})^2} .$$

Moreover, the comparison between high-p_T and low-p_T jets, using the renormalized variable $p_t/\langle p_t \rangle$ should have shown no difference.

However, a hint of the great interest of these studies can be given from the available data on the transverse physics at the ISR.

Figure 37 shows the comparison in the longitudinal momentum distribution between our data and the AFS data on high-p_T jets at ISR [25]. The comparison is made at equivalent energies available for particle production

$$\sqrt{(q_{tot}^{had})^2} = \langle W' \rangle$$

where $\langle W' \rangle$ is the high-p_T jet invariant mass as measured by the AFS collaboration.

The comparison is once again sucessful.

7.   EXPERIMENTAL RESULTS: COMPARISON WITH COLLIDER PHYSICS

In the last paragraph I have outlined the importance of comparing for equal values of

$$\sqrt{(q_{tot}^{had})^2}$$

the multihadronic final states produced at low-p_T and at high-p_T. The closure of the ISR did not allow us to further investigate this subject. But, what was not possible at the ISR could be done at the CERN ($\bar{p}p$) Collider.

An extrapolation of our method to the CERN $\bar{p}p$ Collider is possible because there are two ways of determining

$$\sqrt{(q_{tot}^{had})^2}$$

- one is using the two "leading" particles (this is what we have done in the low-p_T region at the ISR)
- the other, when the "leading" method is out of experimental reach,

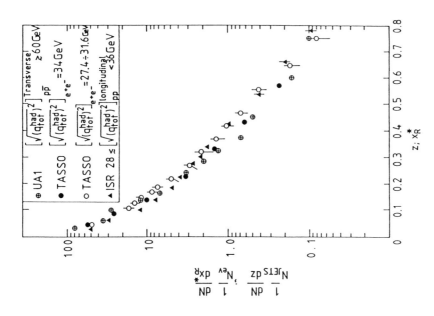

Fig. 38 Comparison, at equivalent $\sqrt{(q_{tot}^{had})^2}$, of the longitudinal momentum distributions for particle produced in pp (ISR low-p_T physics), $p\bar{p}$ (collider high-p_T physics) and e^+e^- (PETRA).

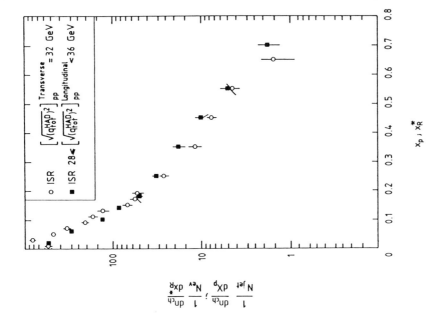

Fig. 37 Longitudinal distribution for particles belonging to high-p_T and low-p_T jets in pp collision at ISR.

is using the measurement of the total energy carried by the two
back-to-back jets (this has been done at the ISR and the Sp$\bar{p}$S in
the high-p_T region). In fact, the basic quantity in comparing
multihadronic states is the effective energy available for particle production, not the nominal.

In the case of two back-to-back jets, it is:

$$\sqrt{(q_{tot}^{had})^2} = \left(\sum_{i=1}^{n} E_i \right)_{jet_1} + \left(\sum_{i=1}^{m} E_i \right)_{jet_2}$$

where E_i is the energy of particle "i" in each of the two back-to-
back jets, whose multiplicities are n for jet_1 and m for jet_2.

There is an important "experimental" difference between the two
ways of determining

$$\sqrt{(q_{tot}^{had})^2} \ .$$

The leading method is straightforward. It only needs the
measurement of the two leading particles.

The other method needs the measurements of all particles in
the two jets, including the neutral ones.

The remarkably unique range of effective hadronic energies
available at the Collider allows a large energy jump from ISR
energies and would produce clear evidence for or against our expectations based on the results obtained at the ISR.

Let us consider an example: if multiparticle hadronic states at
the ($\bar{p}p$) collider are produced at high-p_T with

$$\sqrt{(q_{tot}^{had})} = 100 \text{ GeV} \ .$$

they should be very similar to the multilhadronic systems produced
in (e^+e^-) annihilations, with

$$(\sqrt{s})_{e^+e^-} = 100 \text{ GeV} \ .$$

Such an (e^+e^-) machine is still not available.

What exists and is available are the low-p_T and the high-p_T jets with

$$\sqrt{(q_{tot}^{had})^2} = 100 \text{ GeV}$$

produced at the Collider.

Before going to the experimental results, let me summarize the main ideas of this introduction [24].

To compare multihadronic final states with the same value of

$$\sqrt{(q_{tot}^{had})^2}$$

at high-p_T and low-p_T is of great value to understand the dynamics of strong interactions.

From our findings at the ISR the only difference between high-p_T and low-p_T jets should be the value of $\langle p_t^2 \rangle$, and the difference should disappear if the variable $p_t/\langle p_t \rangle$ is used.

Moreover, at the $(\bar{p}p)$ Collider the slope of

$$\langle p_t \rangle \text{ vs } \sqrt{(q_{tot}^{had})^2}$$

can be measured up to a few hundred GeV. Then we must wait until LEP and SUPERLEP will allow to check these analogies.

7.1 "Transverse" Physics at the CERN $(\bar{p}p)$ Collider

Figure 38 shows the fractional energy distribution measured by the UA1 collaboration at Sp$\bar{p}$S, and by the TASSO collaboration at PETRA, compared with our data taken at the ISR (the scale breaking effects related to the different $\sqrt{(q_{tot}^{had})^2}$ intervals used are small [20] and the comparison is meaningfull). Notice the perfect agreement we obtain once we have subtracted the leading protons.

7.2 Leading effect at Spp̄S

It is reasonable at this point to look for the leading effect at the Spp̄S. To do this, we will extrapolate the ISR results at the Collider energy.

7.2.1 $d\sigma/dx_F$ of the proton

We have tried to find out the $d\sigma/dx_F$ of the proton at the Collider making a Monte Carlo simulation [26] where the inputs were:

i) the best fit to $\langle n_{ch} \rangle^{ISR}$ $\left[\surd(q_{tot}^{had})^2 \right]$ (Fig. 39)

ii) the $\langle n_{ch} \rangle$ measured at the Collider.

The Monte Carlo is based on the idea that $(\surd s)_{pp} \equiv$ superposition of $\surd(q_{tot}^{had})^2$ (see paragraph 1.2.4): to simulate what is happening at a certain $(\surd s)_{pp}$ you need only to generate 2 leading protons according to a certain $d\sigma/dx_F$, calculate for each event $\surd(q_{tot}^{had})^2$ and extract the related multiplicity.

The results of this simulation are shown in Fig. 40. Each slope for the $d\sigma/dx_F$ gives a different prediction for the Collider charged multiplicity. The best prediction is given by the almost flat continuous line which represents the best fit of the ISR data.

Table 1

Results

Slope of leading effect	χ^2/point to $\langle n_{ch} \rangle$ vs $(\surd s)_{pp}$	
Horizontal (Best fit)	1.1	1.1
Decreasing	2.5	2.6
More decreasing	4.4	5.9
Increasing	6.8	6.6
	Without Spp̄S point	With the Spp̄S point

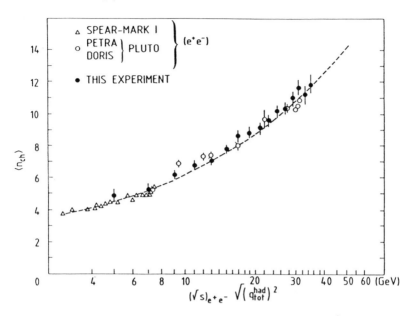

Fig. 39   Mean charged multiplicity as measured in e^+e^- and pp (using the leading proton subtraction).

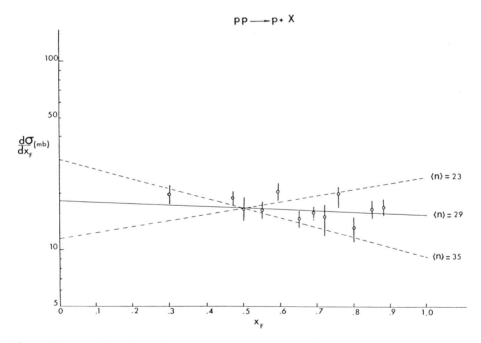

Fig. 40   Predicted UA5 mean charged multiplicity as obtained with various assumption on the $d\sigma/dx_F$. Also shown are the experimental measurements.

Another way to come at the same conclusion is to make a fit of $\langle n_{ch}\rangle$ vs $(\sqrt{s})_{pp}$, comparing the obtained χ^2 for different slopes of the leading effect. The results are shown in Table 1. The horizontal slope is the most appropriate choice to make and for this case we report in Fig. 41 the fit to the charged multiplicity.

7.2.2 Forward-Backward Correlation

Let me remind you the problem [27].

We define the number of charged particles in each hemisphere

$$(n_{ch})_F \qquad \text{and} \qquad (n_{ch})_B \quad .$$

When we measure these quantities, if there is a correlation, we can define its "strength" as

$$\alpha = \frac{d\langle n_{ch}\rangle_F}{d(n_{ch})_B}$$

this "strength" has been proven to increase logarithmically with $(\sqrt{s})_{pp/\bar{p}p}$ from SPS (Fermilab) to ISR up to $(\bar{p}p)$ Collider energies.

Now we want to prove that the behaviour of this strength can be understood if a leading protons effect is present at the Spp̄S. To prove it, even in this case, we have performed a Monte Carlo simulation with the following inputs:

- best fit to the leading effect (Fig. 42)
- best fit to

$$\langle n_{ch}\rangle^{ISR}\left[\sqrt{(q_{tot}^{had})^2}\right] \equiv \langle n_{ch}\rangle_{e^+e^-} \qquad \text{(Fig. 39)}.$$

Our Monte Carlo prediction fits very well with the Spp̄S measurements (Fig. 43).

END OF A MYTH: HIGH-p$_T$ PHYSICS   43

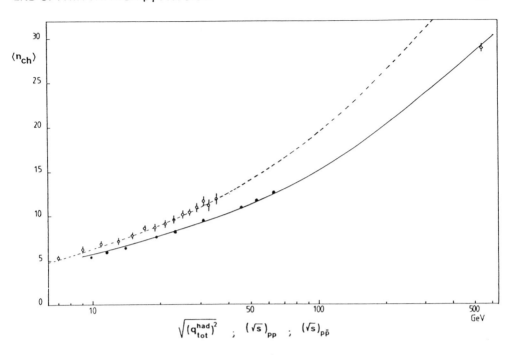

Fig. 41 Dotted line: best fit to e^+e^- and pp (leading subtracted) data. Continuous line: pp mean charged multiplicities as evaluated by Monte Carlo (see text). Also shown are the experimental measurement.

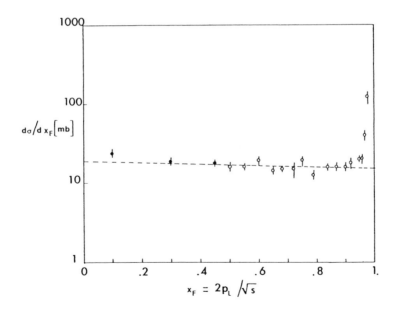

Fig. 42 Best fit to the leading proton effect.

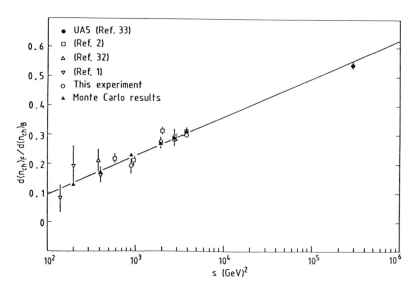

Fig. 43 Comparison of our prediction with the 'correlation-strength' $\alpha = d\langle n_{ch} \rangle_F / d(n_{ch})_B$ measured, in (pp) and (p$\bar{p}$) interactions as a function of s. The full line is the best fit of our Monte Carlo results.

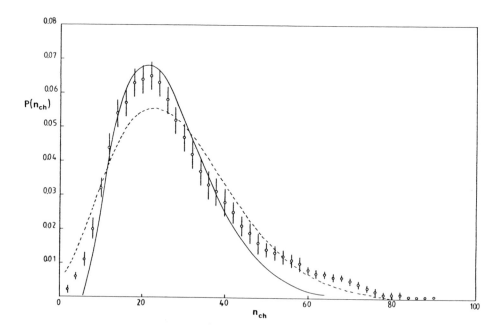

Fig. 44 The charged particle multiplicity distribution measured by the UA5 Collaboration at $(\sqrt{s})_{p\bar{p}} = 540$ GeV. The dashed curve is the KNO scaling prediction, the solid curve the UA5 fit.

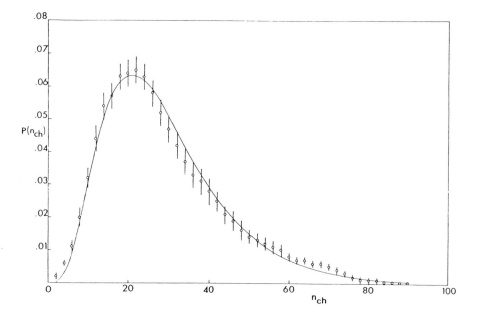

Fig. 45 Our Monte Carlo multiplicity distribution (see text)
compared with UA5 results.

7.2.3 Multiplicity distribution

Until now all attempts to reasonably fit the multiplicity distribution measured by UA5 has failed (Fig. 44) [26].

With our new approach [26] we have tried to solve this problem making a Monte Carlo simulation with the following inputs:

  i) best fit to the leading effect (Fig. 42)

 ii) best fit to $\langle n_{ch}\rangle^{ISR}$ $\left[\mathcal{I}(q_{tot}^{had})^2 \right]$ (Fig. 39)

iii) best fit to the multiplicity distribution in the variable $\mathcal{I}(q_{tot}^{had})^2$ (Fig. 11).

The results are shown in Fig. 45. The impressive agreement confirms once more the importance of the leading effect even at the Collider energy.

7.2.4 Conclusions

From the detailed analysis I have reported we can make the following statements.

a) The leading effect at $Sp\bar{p}S$ is predicted to be very much like that measured at the ISR.

b) Our new method of studying (pp) or (p$\bar{p}$) interactions allows to predict the properties of the multibody hadron states produced at the Collider.

8.   CONCLUSIONS

Multihadronic final states can be produced in :

1) Lepton-hadron processes which can be either

   - <u>weak</u>, such as (vp)

    or

   - <u>e.m.</u>, such as (μp), (ep).

2) Purely e.m. interactions such as (e^+e^-).

3) Purely hadronic interactions such as (pp).

The key point is that there are universality features in these (so far considered) different ways of producing multiparticle hadronic final states.

From the analysis reported we can conclude that:

a) The leading effect must be subtracted and the correct variables must be used if we want to compare purely hadronic processes with (e^+e^-) and DIS.

b) The old myth based on the belief that in order to compare (pp) with (e^+e^-) and DIS you need high-p$_T$(pp) interactions is over.

c) So far we have established a common basis for comparing various ways of producing multiparticle hadronic states. This basis is essential in order to investigate the differences.

d) This new method of studying multihadronic final states should be extended to all possible production processes.

Final conclusion

No matter if the interactions is

strong,

e.m.,

or   weak

the basic quantity which governs the universality features of the multihadronic states is the relativistic invariant quantity:

$$\sqrt{(q_{tot}^{had})^2} \, .$$

REFERENCES

[1]   See lecture by Prof. A. Ali at this school.
[2]   M. Basile et al., Nuovo Cimento 66A, 129, 1981.
[3]   M. Basile et al., Nuovo Cimento Letters 32, 321, 1981.
[4]   M. Basile et al., Nuovo Cimento 79A, 1, 1983.
[5]   M. Basile et al., Nuovo Cimento 58A, 193, 1980.
[6]   M. Basile et al., Physics Letters 92B, 367, 1980.
[7]   M. Basile et al., Nuovo Cimento 65A, 414, 1981.
[8]   M. Basile et al., Nuovo Cimento 67A, 53, 1982.
[9]   M. Basile et al., Nuovo Cimento 95A, 311, 1980.
[10] M. Basile et al., Nuovo Cimento 65A, 400, 1981.
[11] M. Basile et al., Nuovo Cimento 67A, 244, 1982.
[12] M. Basile et al., Preprint CERN-EP/84-94, 1984.

48   A. ZICHICHI ET AL.

[13] M. Basile et al., Physics Letters 99B, 247, 1981.
[14] M. Basile et al., Nuovo Cimento Letters 32, 210, 1981.
[15] M. Basile et al., Nuovo Cimento Letters 31, 273, 1981.
[16] M. Basile et al., Nuovo Cimento Letters 29, 491, 1980.
[17] M. Basile et al., Nuovo Cimento 73A, 329, 1983.
[18] G. Bonvicini et al., Nuovo Cimento Letters 36, 563, 1983.
[19] J. Berbiers et al., Nuovo Cimento Letters 37, 246, 1983.
[20] M. Basile et al., Nuovo Cimento Letters 38, 289, 1983.
[21] M. Basile et al., Nuovo Cimento Letters 36, 303, 1983.
[22] G. Bonvicini et al., Nuovo Cimento Letters 36, 555, 1983.
[23] G. Bonvicini et al., Nuovo Cimento Letters 37, 289, 1983.
[24] M. Basile et al., Nuovo Cimento Letters 38, 367, 1983.
[25] T. Åkesson, Preprint CERN-EP/83-130, 1983.
[26] M. Basile et al., Preprint CERN-EP/84-95, 1984.
[27] M. Basile et al., Nuovo Cimento Letters 38, 359, 1983.

DISCUSSION

Chairman: A. Zichichi

Scientific Secretary: R. Nania

- VAFA

In identifying the leading particles there is no difficulty when $\sqrt{s} \gg \sqrt{\left(q_{tot}^{HAD}\right)^2}$. Problems may arise when $\sqrt{s} \gtrsim \sqrt{\left(q_{tot}^{HAD}\right)^2}$. Does the universality become less pronounced for such cases ?

- ZICHICHI

No. Universality ought to remain for any value of $\sqrt{\left(q_{tot}^{HAD}\right)^2}$. In our case, the leading particle subtraction has been carried out and universality checked when the proton carries a fraction of the initial momentum in the range ($.4 < x_F < .8$). Since we identify a proton as the fastest particle in the hemisphere, we are sure that in this region pion contamination is low. To check universality at small x_F we would need Cherenkov counters or TOF systems, but such an experiment has not been possible because of lack of ISR machine time.

- VAFA

But would you be able to identify which is the leading particle in such a low x_F region ?

- ZICHICHI

The baryons centrally produced will, of course, contaminate the sample, but, nevertheless, the analysis should be possible.

- *LIM*

    The leading quantity L is defined as the ratio of two finite integrals. Is the definition of L limit independent ?

- *ZICHICHI*

    As already pointed out in my lecture, the limits chosen are purely conventional: it is easier to remember that three propagating quarks give L = 3. In any case, what is important is not the value L assumes, but the fact that, independently of the process studied, the quantity L remains the same and it gives an idea of how much a particle is leading.

- *CAI XU*

    What is the exact meaning of the leading particle in e^+e^- and high-p_T collisions ?

- *ZICHICHI*

    In my opinion, the key point lies on the probability that the fragmentation products recombine themselves into one leading particle. From high-p_T data at the ISR and the Collider, we have

$$\frac{\sigma(\text{single particle})^{\text{inclusive}}_{E_T}}{\sigma(\text{jet})^{\text{inclusive}}_{E_T}} \approx 10^{-2} \div 10^{-3}$$

for $E_T \simeq 5 \div 10$ GeV; from e^+e^-

$$\frac{\sigma(e^+e^- \to D^* \ X)}{\sigma(e^+e^- \to c \ X)} \approx 10^{-2}$$

We can deduce that if one single particle must carry a great part of a certain energy available, it should pay a factor $\sim 10^{-2}$. How this recombination into one leading particle can be possible is a question for QCD theorists.

- *CAI XU*

The Collider data on the multiplicity distribution refer only to particles in the central region, while KNO scaling is valid on the full rapidity region. It is meaningless to use the KNO scaling function for their data.

- *ZICHICHI*

The UA5 data, which we have used, are in the full rapidity region. Even our multiplicity distributions as function of $\sqrt{\left(\frac{HAD}{q_{tot}}\right)^2}$ are in the full rapidity range, and, as we have shown, they scale. But while the KNO function doesn't succeed in explaining the UA5 data, we do.

- *HARNEW*

Do you actually trigger on protons ?   If not, how can you identify them ?

- *ZICHICHI*

To identify the protons we use the fact that in the region $.4 < x_F < .8$ the p/π ratio increases from 2 up to ~20; so, taking the fastest particle in the hemisphere with x_F in this specific range, the π contamination of our sample of leading protons goes from 30% to 5%.

- *HARNEW*

So if you look at the diffractive part, you will normally lose the fastest down the beam pipe.

- *ZICHICHI*

This is not correct because the diffractive part is excluded with the cut at $x_F < 0.8$.

- *HARNEW*

UA2 has shown data on $\langle n_{ch} \rangle$ versus $m_{jet-jet}$ where they have agreement with TASSO only at the higher energies, while at low $m_{jet-jet}$ they considerably deviate. Can you comment on this ?

- *ZICHICHI*

The data were preliminary and, in the deviation region, almost flat. It could be that they have evidence for gluon induced jets, but I think it is too early to make any comment.

- *PICCIONI*

You have shown a correlation in the forward-backward charged multiplicity. It would be nice to see if, plotting the distribution of the difference of the multiplicities, such distribution would appear as approximately a Gaussian with a width equal $\sqrt{2}$ times the width of each distribution. This for the same effective energy.

- *ZICHICHI*

We can try to do this, but it doesn't seem to me that we can gain new information.

- *DI LELLA*

You have shown that, using your new approach, the way hadronization takes place does not depend on the different processes involved. But people who have heard you should not have the impression that high-p_T physics is useless. High-p_T at the Collider does not address any question about hadronization. High-p_T physics studies essentially the scattering between the partons in a region where QCD is simple because it is asymptotically free. In fact the study has concentrated on the p_T and the angular distributions of these jets, and not on their hadronization, which at this stage is only a complication.

- *ZICHICHI*

There are two myths: 1) in order to compare multibody final states you can only use large p_T hadronic processes; 2) in order to understand QCD you should only use large p_T processes. The first myth is over. The second is in good shape, as far as mythology is concerned. In fact QCD ought to explain the leading effect and all hadronic processes if it has to be a real theory, not a model with many parameters ad hoc.

- *KOMAMIYA*

Jets from e^+e^- are essentially quark jets; jets from pp are

mostly gluon jets. If one can compare these hadronic jets by re-
scaling the c.m. energy, is it possible to see the difference
between quark-jets and gluon-jets (for example the p_t broadening of
gluon jets) ?.

- *ZICHICHI*

Probably the discrepancies we find in the $n_{ch}/<n_{ch}>$ distributions or in the p_t^2 can find an explanation in this sense. Since in
pp interactions gluon jets tend to be in large $\cos\theta$ regions, we have
proposed to compare forward jets and transverse jets. So far we
have not been able to convince the committee that this is worth
studying.

- *ROHM*

In your lecture, you concentrated on the statistic of events
with a given effective hadronic energy. How does the probability
of obtaining a given hadronic energy scale with the incident particle energy ?

- *ZICHICHI*

This probability is given by the $d\sigma/dx_F$ distribution of the
protons. As this scales at ISR energies, it is expected also to
scale at Collider energies. In fact, we have indirectly proved that
this is so. See our multiplicity distribution and the forward backward multiplicity correlation predictions.

- *BONINO*

What is the leading rate in the various processes you consider ?
If there is a difference, do you understand the reason ?

- *ZICHICHI*

In pp about 20 mb of cross-section are in the leading effect.
In e^+e^- about 1% of the total cross-section. For the explanation
you must wait until QCD wakes up.

498

Universality Features in Multihadronic Final States[(*)]

A. Zichichi

INFN and University of Bologna, Italy

CERN - Geneva, Switzerland

This year I was scheduled for a Lecture but I shall instead talk only ten minutes to allow for another Lecture to be included in the programme of the School. The topic of my Lecture was relative to a set of experimental data which is many years old. Nevertheless, these data appear to me to be very interesting. In fact, no matter if the interactions are strong, electromagnetic or weak, if these interactions produce multihadronic final states with the same "Effective Energy", the properties of these final states are identical. By properties I mean, for example, the distributions of the average charged multiplicity,

$$< n_{ch} >,$$

of the fractional energy

$$\frac{d\sigma}{dx_i},$$

of the fractional transverse momentum

$$\frac{d\sigma}{d\sigma_{T_i}},$$

the ratio of "charged" to "total" energy, the "planarity" of the multihadronic systems produced, etc .... .

There is however a difficulty: being the author (together with my collaborators) of these findings, I can obviously overestimate their importance. The proof is that these results have been presented on several occasions [1]; the reason being simply because I want to bring it to the attention of those who

---

[(*)] Old but still interesting results on the properties of multihadronic final states produced in Strong, Electromagnetic and Weak Interactions: a brief recalling.

work in QCD. In fact, the Universality Features are a QCD non perturbative effect and, as Professor Gribov said to me on several occasions [2], it will be possible to understand them when we are able to have a QCD theory which includes simultaneously perturbative and non perturbative phenomena.

For this year I will limit myself to invite the distinguished QCD specialists here present to give some thought to our results and let me know their reactions.

Please do not forget that the BCF collaboration made a systematic study of the multihadronic final states produced in low$-p_T$ (pp) interactions at the ISR and compared the results with those obtained in the processes listed below:

Process	Data Sources
(e^+e^-)	SLAC, DORIS, PETRA
(DIS)	SPS/EMC
(pp) ⎤ Transverse physics ⎡	ISR (AFS)
$(\bar{p}p)$ ⎦	SPS Collider (UA1)
(e^+e^-)	PETRA/TASSO (leading effect in D* production).

The results of these studies [3-18] show that, once a common basis for comparison is found by the use of the correct variables, remarkable analogies are observed in processes so far considered basically different like

- low$-p_T$ (pp) interactions
- (e^+e^-) annihilations
- (DIS) processes
- high$-p_T$ (pp) and $(\bar{p}p)$ interactions.

In order to establish a common basis for a detailed comparison[**] of the multihadronic final states, it is crucial to identify the "Effective Energy".

[**] The root of this new approach to the study of hadronic interactions goes back a long time to a proposal by the CERN-Bologna group: "Study of deep inelastic high momentum transfer hadronic collisions" PMI/com-69/35, 8 July 1969."

500

In this way, the Universality Features of the multihadronic final states produced in Strong, Electromagnetic and Weak Interactions show up. These Universality Features are reported in a set of graphs which, in order to save time, I will not show here. They are all at your disposal in the secretariat of the School.

And now, as it is the tradition of the School for the morning sessions, only very urgent questions or remarks are allowed.

Morning Discussion Session:

V. Gribov: I have three remarks.

Remark n. 1: When I read the paper "Evidence of the same multiparticle production mechanism in (pp) collisions as in (e^+e^-) annihilation", I realized that something very interesting had been found. In fact the introduction of the "Effective Energy" in the analysis of (pp) collisions at the CERN-ISR gave a totally unexpected result.

Remark n. 2: In the physics community there was a sort of gentlemen's agreement: please do not speak about results in contrast with the so much searched for gauge interaction to describe hadronic phenomena.

These "hidden" results were the hadronic systems produced in the interactions between pairs of hadrons; they were all different. Each pair of interacting particles, when producing systems consisting of many hadronic particles, had its own final state. No-one knew how to settle this flagrant contradiction. I wish I had the idea of the "Effective Energy".

Remark n. 3: Think of it. Even after so many years it appears to me a great achievement in physics.

A. Zichichi: Thank you, Volodya. I have a question. Why QCD is not able to "predict" the Universality Features?

V. Gribov: As you have emphasised in your lecture this is a non perturbative QCD problem. In order to answer your question I first need to understand confinement.

501

References

[1]   *WHAT WE CAN LEARN FROM HIGH-ENERGY, SOFT (pp) INTERACTIONS*, Proceedings of the XIX Course of the "Ettore Majorana" International School of Subnuclear Physics, Erice, Italy, 31 July-11 August 1981: "The Unity of the Fundamental Interactions" (Plenum Press, New York-London, 1983), 695; *HIGH-ENERGY SOFT (pp) INTERACTIONS COMPARED WITH (e^+e^-) AND DEEP INELASTIC SCATTERING*, Proceedings of the XX Course of the "Ettore Majorana" International School of Subnuclear Physics, Erice, Italy, 3-14 August 1982: "Gauge Interactions: Theory and Experiment" (Plenum Press, New York-London, 1984), 701; *HADRON COLLIDERS VERSUS (e^+e^-) COLLIDERS: A CONTRIBUTION TO THE ROUND TABLE FROM THE BCF GROUP*, Proceedings of the 3rd Topical Workshop on "Proton-Antiproton Collider Physics", Rome, Italy, 12-14 January 1983, CERN 83-04, 409; *STUDY OF SOFT (pp) INTERACTIONS AND COMPARISON WITH (e^+e^-) AND DIS*, Proceedings of the XVIII Rencontre de Moriond on "Gluons and Heavy Flavours", La Plagne, France, 23-29 January 1983 (Ed. Frontières, Gif-sur-Yvette, 1983), 175; *THE END OF A MYTH: HIGH-P_T PHYSICS*, Opening Lecture in Proceedings of the XXII Course of the "Ettore Majorana" International School of Subnuclear Physics, Erice, Italy, 5-15 August 1984: "Quarks, Leptons, and their Constituents" (Plenum Press, New York-London, 1988), 1; *UNIVERSALITY PROPERTIES IN NON-PERTURBATIVE QCD - LEADING IN (e^+e^-) D* AT PETRA*, Proceedings of the XXIII Course of the "Ettore Majorana" International School of Subnuclear Physics, Erice, Italy, 4-14 August 1985: "Old and New Forces of Nature" (Plenum Press, New York-London, 1988), 117; *LEADING HEAVY FLAVOURED BARYON PRODUCTION AT ISR*, Proceedings of the 10th Warsaw Symposium on "Elementary Particle Physics", Kazimierzs, Poland, 24-30 May 1987 (Warsaw Univ. and INS, Warsaw, 1987), 237; *THE INFN ELOISATRON PROJECT*, Proceedings of the HARC '93 International Workshop on "Recent Advances in the Superworld", Houston Advanced Research Center, The Woodlands, TX, USA, 14-16 April 1993 (World Scientific, 1994), 363; *"LEADING" PHYSICS AT LHC INCLUDING MACHINE STUDIES PLUS DETECTOR R&D (LAA)*, Proceedings of the XXXIII Course of the "Ettore Majorana" International School of Subnuclear Physics, Erice, Italy, 2-10 July 1995: "Vacuum and Vacua: The Physics of Nothing" (World Scientific, 1995), 381.

[2]   *INFN ELOISATRON PROJECT*, 17th Workshop: *"QCD at 200 TeV"*, (11-17 June 1991); *INTERNATIONAL SCHOOL OF SUBNUCLEAR PHYSICS*, 33rd Course: *"Vacuum and Vacua: The Physics of Nothing"*, (2-10 July 1995).

[3]   *THE "LEADING"-PARTICLE EFFECT IN HADRON PHYSICS*, Nuovo Cimento <u>66A</u>, 129 (1981).

[4]   *THE "LEADING"-BARYON EFFECT IN STRONG, WEAK, AND ELECTROMAGNETIC INTERACTIONS*, Lettere al Nuovo Cimento <u>32</u>, 321 (1981).

[5]   *UNIVERSALITY FEATURES IN (pp), (e^+e^-) AND DEEP-INELASTIC-SCATTERING PROCESSES*, Nuovo Cimento <u>79A</u>, 1 (1984).

502

[6] *THE FRACTIONAL MOMENTUM DISTRIBUTION IN p-p COLLISIONS COMPARED WITH e^+e^- ANNIHILATION*, Nuovo Cimento 58A, 193 (1980).

[7] *EVIDENCE OF THE SAME MULTIPARTICLE PRODUCTION MECHANISM IN p-p COLLISIONS AS IN e^+e^- ANNIHILATION*, Physics Letters 92B, 367 (1980).

[8] *CHARGED-PARTICLE MULTIPLICITIES IN (pp) INTERACTIONS AND COMPARISON WITH (e^+e^-) DATA*, Nuovo Cimento 65A, 400 (1981).

[9] *THE INCLUSIVE MOMENTUM DISTRIBUTION IN (pp) REACTIONS, COMPARED WITH LOW-ENERGY (e^+e^-) DATA IN THE RANGE $(\sqrt{s})_{e^+e^-} = (3.0\text{-}7.8)$ GeV*, Nuovo Cimento 67A, 53 (1982).

[10] *THE ENERGY DEPENDENCE OF CHARGED PARTICLE MULTIPLICITY IN p-p INTERACTIONS*, Physics Letters 95B, 311 (1980).

[11] *CHARGED-PARTICLE MULTIPLICITIES IN (pp) INTERACTIONS AND COMPARISON WITH (e^+e^-) DATA*, Nuovo Cimento 65A, 400 (1981).

[12] *A DETAILED STUDY OF $\langle n_{ch}\rangle$ VERSUS E^{had} AND $m_{1,2}$ AT DIFFERENT $(\sqrt{s})_{pp}$ IN (pp) INTERACTIONS*, Nuovo Cimento 67A, 244 (1982).

[13] *THE RATIO OF CHARGED-TO-TOTAL ENERGY IN HIGH-ENERGY PROTON-PROTON INTERACTIONS*, Physics Letters 99B, 247 (1981).

[14] *THE INCLUSIVE TRANSVERSE-MOMENTUM DISTRIBUTION IN HADRONIC SYSTEMS PRODUCED IN PROTON-PROTON COLLISIONS*, Lettere al Nuovo Cimento 32, 210 (1981).

[15] *THE TRANSVERSE-MOMENTUM DISTRIBUTIONS OF PARTICLES PRODUCED IN pp REACTIONS AND COMPARISON WITH e^+e^-*, Lettere al Nuovo Cimento 31, 273 (1981).

[16] *MEASUREMENTS OF $\langle p_T^2\rangle_{in}$ AND $\langle p_T^2\rangle_{out}$ DISTRIBUTIONS IN HIGH-ENERGY pp INTERACTIONS*, Lettere al Nuovo Cimento 29, 491 (1980).

[17] *EXPERIMENTAL PROOF THAT THE LEADING PROTONS ARE NOT CORRELATED*, Nuovo Cimento 73A, 329 (1983).

[18] *EVIDENCE FOR THE SAME TWO-PARTICLE CORRELATIONS IN RAPIDITY SPACE IN (pp) COLLISIONS AND (e^+e^-) ANNIHILATION*, Lettere al Nuovo Cimento 36, 563 (1983).

IL NUOVO CIMENTO   VOL. 67 A, N. 3   1 Febbraio 1982

A Detailed Study of $\langle n_{\mathrm{ch}} \rangle$ vs. E^{had} and $m_{1,2}$ at Different $(\sqrt{s})_{\mathrm{pp}}$ in (pp) Interactions.

M. Basile, G. Bonvicini, G. Cara Romeo, L. Cifarelli, A. Contin,
M. Curatolo, G. D'Alí, P. Di Cesare, B. Esposito, P. Giusti,
T. Massam, R. Nania, F. Palmonari, A. Petrosino, V. Rossi,
G. Sartorelli, M. Spinetti, G. Susinno, G. Valenti,
L. Votano and A. Zichichi

CERN - Geneva, Switzerland
Istituto di Fisica dell'Università - Bologna, Italia
Istituto Nazionale di Fisica Nucleare - Laboratori Nazionali di Frascati, Italia
Istituto Nazionale di Fisica Nucleare - Sezione di Bologna, Italia
Istituto di Fisica dell'Università - Perugia, Italia
Istituto di Fisica dell'Università - Roma, Italia

(ricevuto il 23 Novembre 1981)

Summary. — By using (pp) interactions at three different c.m. energies, $(\sqrt{s})_{\mathrm{pp}} = 30, 44, 62$ GeV, it is shown that the average charged-particle multiplicity $\langle n_{\mathrm{ch}} \rangle$ vs. the invariant mass of the hadronic system $m_{1,2}$ has the same behaviour as it has vs. $2E^{\mathrm{had}}$. Moreover, in both cases $\langle n_{\mathrm{ch}} \rangle$ is shown to be nearly independent of $(\sqrt{s})_{\mathrm{pp}}$ and in good agreement with the average charged-particle multiplicity measured in the (e⁺e⁻) annihilation.

1. – Introduction and purpose of the experiment.

We have already reported on a measurement of the average charged-particle multiplicity $\langle n_{\mathrm{ch}} \rangle$ in (pp) interactions at $(\sqrt{s})_{\mathrm{pp}} = 30, 44$ and 62 GeV total c.m. energies [1]. The value of $\langle n_{\mathrm{ch}} \rangle$ was measured as a function of E^{had},

---

[1] M. Basile, G. Cara Romeo, L. Cifarelli, A. Contin, G. D'Alí, P. Di Cesare,

A DETAILED STUDY ETC. **245**

the energy available for particle production, once the energy carried away by the «leading» outgoing proton is subtracted. For fixed values of E^{had}, $\langle n_{ch} \rangle$ was found to be independent of $(\sqrt{s})_{pp}$. Moreover, the behaviour of $\langle n_{ch} \rangle$ vs. $2E^{had}$ was found to be in good agreement with the results obtained in $(e^+ e^-)$ annihilation, when $2E^{had} = (\sqrt{s})_{e^+ e^-}$.

In our analysis of (pp) collisions we have already introduced ([2]) the quantity $m_{1,2}$, which is

$$(1) \qquad m_{1,2} = [(E_1^{had} + E_2^{had})^2 - (\boldsymbol{p}_1^{had} + \boldsymbol{p}_2^{had})^2]^{\frac{1}{2}},$$

where E_1^{had}, $\boldsymbol{p}_1^{had}$ and E_2^{had}, $\boldsymbol{p}_2^{had}$ are the energy and momentum differences between the incident protons and the outgoing leading protons in the two hemispheres ([3-8]). The quantity $m_{1,2}$ represents the invariant mass of the whole hadronic system which remains once the two outgoing leading protons are subtracted.

The purpose of the present work is to see whether $\langle n_{ch} \rangle$ vs. $m_{1,2}$ has the same behaviour as when it is studied in terms of E^{had}. In this case it had to scale with $(\sqrt{s})_{pp}$ and it had to be in good agreement with $e^+ e^-$ data when $2E_{beam}^{e^+,e^-} = 2E^{had} = m_{1,2}$. Moreover, it is important to study whether $\langle n_{ch} \rangle$ vs. $m_{1,2}$ depends on the selection of given values of E^{had}, and also whether $\langle n_{ch} \rangle$ vs. E^{had} depends on any selection of $m_{1,2}$ values.

B. ESPOSITO, P. GIUSTI, T. MASSAM, R. NANIA, F. PALMONARI, V. ROSSI, G. SARTORELLI, M. SPINETTI, G. SUSINNO, G. VALENTI, L. VOTANO and A. ZICHICHI: *Nuovo Cimento A*, **65**, 400 (1981).

(²) M. BASILE, G. CARA ROMEO, L. CIFARELLI, A. CONTIN, G. D'ALÍ, P. DI CESARE, B. ESPOSITO, P. GIUSTI, T. MASSAM, R. NANIA, F. PALMONARI, G. SARTORELLI, M. SPINETTI, G. SUSINNO, G. VALENTI and A. ZICHICHI: *Phys. Lett. B*, **99**, 247 (1981).

(³) M. BASILE, G. CARA ROMEO, L. CIFARELLI, A. CONTIN, G. D'ALÍ, P. DI CESARE, B. ESPOSITO, P. GIUSTI, T. MASSAM, F. PALMONARI, G. SARTORELLI, G. VALENTI and A. ZICHICHI: *Phys. Lett. B*, **92**, 367 (1980).

(⁴) M. BASILE, G. CARA ROMEO, L. CIFARELLI, A. CONTIN, G. D'ALÍ, P. DI CESARE, B. ESPOSITO, P. GIUSTI, T. MASSAM, R. NANIA, F. PALMONARI, G. SARTORELLI, G. VALENTI and A. ZICHICHI: *Phys. Lett. B*, **95**, 311 (1980).

(⁵) M. BASILE, G. CARA ROMEO, L. CIFARELLI, A. CONTIN, G. D'ALÍ, P. DI CESARE, B. ESPOSITO, P. GIUSTI, T. MASSAM, R. NANIA, F. PALMONARI, G. SARTORELLI, G. VALENTI and A. ZICHICHI: *Nuovo Cimento A*, **58**, 193 (1980).

(⁶) M. BASILE, G. CARA ROMEO, L. CIFARELLI, A. CONTIN, G. D'ALÍ, P. DI CESARE, B. ESPOSITO, P. GIUSTI, T. MASSAM, R. NANIA, F. PALMONARI, G. SARTORELLI, G. VALENTI and A. ZICHICHI: *Lett. Nuovo Cimento*, **29**, 491 (1980).

(⁷) M. BASILE, G. CARA ROMEO, L. CIFARELLI, A. CONTIN, G. D'ALÍ, P. DI CESARE, B. ESPOSITO, P. GIUSTI, T. MASSAM, R. NANIA, F. PALMONARI, G. SARTORELLI, M. SPINETTI, G. SUSINNO, G. VALENTI and A. ZICHICHI: *Lett. Nuovo Cimento*, **30**, 389 (1981).

(⁸) M. BASILE, G. CARA ROMEO, L. CIFARELLI, A. CONTIN, G. D'ALÍ, P. DI CESARE, B. ESPOSITO, P. GIUSTI, T. MASSAM, R. NANIA, F. PALMONARI, G. SARTORELLI, G. VALENTI and A. ZICHICHI: *Lett. Nuovo Cimento*, **31**, 273 (1981).

2. – Data analysis and results.

The experiment was done at the CERN Intersecting Storage Rings (ISR) using the Split-Field Magnet (SFM) and its powerful multiwire proportional chamber (MWPC) assembly [9]. For details of the experimental set-up and

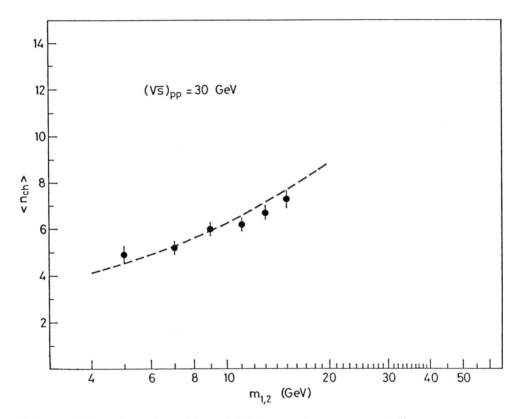

Fig. 1. – Mean charged-particle multiplicity $\langle n_{\text{ch}} \rangle$ vs. $m_{1,2}$ at $(\sqrt{s})_{\text{pp}} = 30$ GeV. The dashed line is our previously published fit [1].

data analysis, we refer the reader to our previous papers [1-8]. The total number of selected events with two leading protons, one in each hemisphere, with x_{F} ($x_{\text{F}} = 2p_{\text{L}}/\sqrt{s}$, where p_{L} is the longitudinal proton momentum) in the range

$$(2) \qquad\qquad 0.42 \leqslant x_{\text{F}} \leqslant 0.86$$

[9]   R. BOUCLIER, R. C. A. BROWN, E. CHESI, L. DUMPS, H. G. FISCHER, P. G. INNO-CENTI, G. MAURIN, A. MINTEN, L. NAUMANN, F. PIUZ and O. ULLALAND: Nucl. Instrum. Methods, 125, 19 (1975).

is 9850, of which 1490 are collected at $(\sqrt{s})_{\mathrm{pp}} = 30$ GeV, 5260 at $(\sqrt{s})_{\mathrm{pp}} = 44$ GeV and 3100 at $(\sqrt{s})_{\mathrm{pp}} = 62$ GeV. In addition to these data, in which the trigger was chosen such that it would enrich the samples with two leading protons, we have also analysed 1150 events at $(\sqrt{s})_{\mathrm{pp}} = 62$ GeV, taken earlier in the « minimum bias » trigger mode (we will call this sample the « old data ») [1].

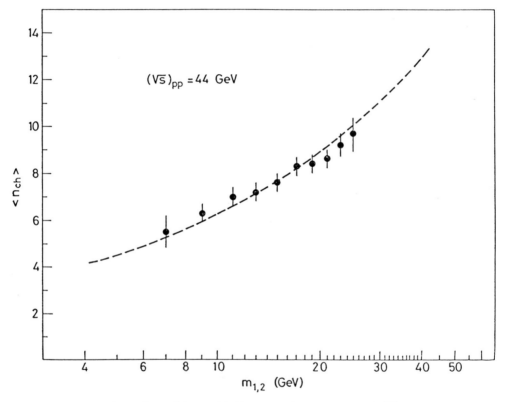

Fig. 2. – Mean charged-particle multiplicity $\langle n_{\mathrm{ch}} \rangle$ vs. $m_{1,2}$ at $(\sqrt{s})_{\mathrm{pp}} = 44$ GeV. The dashed line is our previously published fit [1].

The charged-particle multiplicity has been measured by counting the tracks in the whole event, without any cut in momentum resolution. The observed multiplicities have been corrected for detection efficiency via Monte Carlo simulation. This correction is, on the average, less than 30 %, and introduces a systematic error $\leqslant 5\%$ amongst the four sets of data and an overall systematic uncertainty $\leqslant 8\%$.

The contribution to $\langle n_{\mathrm{ch}} \rangle$ due to $K_{\mathrm{s}}^0 \to \pi^+\pi^-$ has been subtracted. Other small contributions, such as γ conversion, have also been subtracted.

Since protons are not directly identified, there is a contamination from positive pions. In the selected range of x_{F} (eq. (2)) this contamination varies

M. BASILE, G. BONVICINI, G. CARA ROMEO, L. CIFARELLI, A. CONTIN, ETC.

TABLE I. – *Mean charged-particle multiplicity* $\langle n_{ch} \rangle$ *vs.* $2E^{had}$ *and* $m_{1,2}$ *for different* $(\sqrt{s})_{pp}$ *of the* ISR. *In both cases events with two leading protons, one for each hemisphere, have been used. The quoted errors are statistical only. The systematic uncertainty is estimated to be less than 8%.*

$\langle n_{ch} \rangle$ vs. $2E^{had}$					
$2E^{had}$ (GeV)	$(\sqrt{s})_{pp}$				
	30 GeV	44 GeV	62 GeV	62 GeV « old data »	Average
5	4.4 ± 0.3				4.4 ± 0.3
7	5.3 ± 0.3	5.4 ± 0.3			5.4 ± 0.2
9	5.3 ± 0.3	6.3 ± 0.3	6.6 ± 0.5		6.1 ± 0.2
11	6.2 ± 0.3	6.7 ± 0.3	7.5 ± 0.5		6.7 ± 0.2
13	7.6 ± 0.3	7.4 ± 0.3	7.8 ± 0.5	8.1 ± 1.0	7.7 ± 0.2
15	7.1 ± 0.3	7.8 ± 0.3	8.0 ± 0.5		7.6 ± 0.2
17	8.1 ± 0.4	8.2 ± 0.3	9.4 ± 0.5	8.2 ± 1.0	8.4 ± 0.3
19		8.0 ± 0.3	9.4 ± 0.5		8.5 ± 0.3
21		9.0 ± 0.3	9.5 ± 0.5		9.2 ± 0.3
23		8.9 ± 0.3	10.2 ± 0.5	9.0 ± 0.6	9.5 ± 0.3
25		10.0 ± 0.4	10.8 ± 0.5		10.3 ± 0.4
27			11.1 ± 0.5	10.3 ± 0.7	10.9 ± 0.5
29			11.1 ± 0.5		11.0 ± 0.5
31			11.3 ± 0.5		11.1 ± 0.5
33			11.1 ± 0.5	10.8 ± 0.5	11.0 ± 0.5
35			11.7 ± 0.5		11.7 ± 0.5

$\langle n_{ch} \rangle$ vs. $m_{1,2}$					
$m_{1,2}$ (GeV)	$(\sqrt{s})_{pp}$				
	30 GeV	44 GeV	62 GeV	62 GeV « old data »	Average
5	4.9 ± 0.4				4.9 ± 0.4
7	5.2 ± 0.3	5.5 ± 0.7			5.3 ± 0.3
9	6.0 ± 0.3	6.3 ± 0.4			6.2 ± 0.3
11	6.2 ± 0.3	7.0 ± 0.4	7.8 ± 0.8		6.8 ± 0.3
13	6.7 ± 0.3	7.2 ± 0.4	7.5 ± 0.7	7.3 ± 1.0	7.1 ± 0.3
15	7.3 ± 0.4	7.6 ± 0.4	9.1 ± 0.6		7.8 ± 0.3
17		8.3 ± 0.4	8.9 ± 0.5	9.1 ± 0.7	8.6 ± 0.4
19		8.4 ± 0.4	9.4 ± 0.5		8.8 ± 0.4
21		8.6 ± 0.4	9.8 ± 0.5		9.1 ± 0.4
23		9.2 ± 0.5	9.9 ± 0.5	9.9 ± 0.6	9.6 ± 0.4
25		9.7 ± 0.8	10.4 ± 0.4		10.2 ± 0.4
27			10.6 ± 0.4	9.8 ± 0.6	10.4 ± 0.4
29			11.1 ± 0.5		11.0 ± 0.5
31			11.8 ± 0.5		11.7 ± 0.5
33			11.2 ± 0.6	10.7 ± 0.7	11.2 ± 0.6
35			11.9 ± 0.6		11.9 ± 0.6

from 25 % to 2 % [10]. This misidentification has been studied via a p_T-limited phase-space Monte Carlo and produces a change in $\langle n_{ch} \rangle$ of about 0.5 charged unit at the highest $m_{1,2}$.

In table I and in fig. 1 to 4, the values of $\langle n_{ch} \rangle$ as a function of $m_{1,2}$ are reported for the three values $(\sqrt{s})_{pp} = 30$, 44 and 62 GeV. There is a good

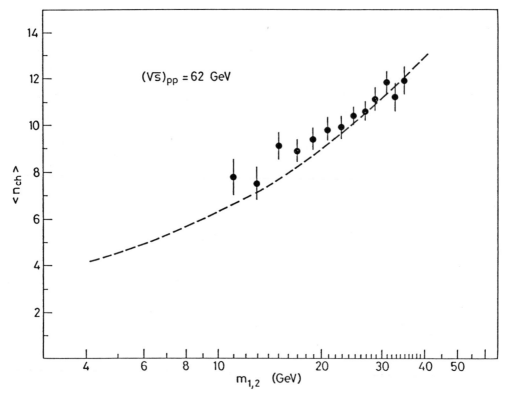

Fig. 3. – Mean charged-particle multiplicity $\langle n_{ch} \rangle$ vs. $m_{1,2}$ at $(\sqrt{s})_{pp} = 62$ GeV. The dashed line is our previously published fit [1].

agreement with our previously published [1] best fit to $\langle n_{ch} \rangle$ vs. E^{had} (dashed line). These data show that, within the experimental uncertainties, $\langle n_{ch} \rangle$ vs. $m_{1,2}$ nearly scales with $(\sqrt{s})_{pp}$. In table I the values of $\langle n_{ch} \rangle$, averaged at a given value of $m_{1,2}$ or $2E^{had}$ over the different $(\sqrt{s})_{pp}$ samples, are reported. These data are shown in fig. 5 and 6. The agreement with the e^+e^- data, as

[10] P. CAPILUPPI, G. GIACOMELLI, A. M. ROSSI, G. VANNINI and A. BUSSIÈRE: *Nucl. Phys. B*, **70**, 1 (1974); P. CAPILUPPI, G. GIACOMELLI, A. M. ROSSI, G. VANNINI, A. BERTIN, A. BUSSIÈRE and R. J. ELLIS: *Nucl. Phys. B*, **79**, 189 (1974).

M. BASILE, G. BONVICINI, G. CARA ROMEO, L. CIFARELLI, A. CONTIN, ETC.

well as with our already published data ([1]), is excellent. This can be deduced from the dashed lines which simultaneously fit our data and the (e^+e^-) data.

We now report on the detailed study of $\langle n_{ch} \rangle$ *vs.* $m_{1,2}$ and $2E^{had}$.

As a first test we have studied the influence of the quantity E^{had} on $\langle n_{ch} \rangle$ *vs.* $m_{1,2}$. For this purpose, two samples of events were selected:

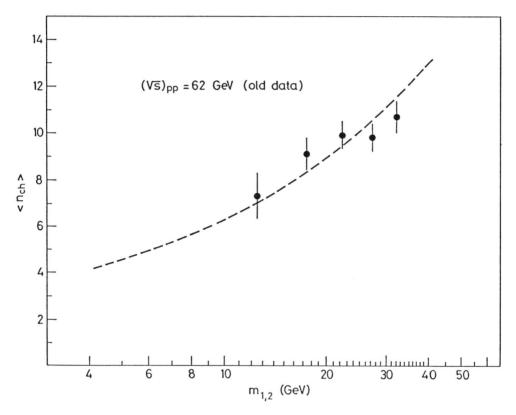

Fig. 4. – Mean charged-particle multiplicity $\langle n_{ch} \rangle$ *vs.* $m_{1,2}$ at $(\sqrt{s})_{pp} = 62$ GeV by using the « old data » sample. The dashed line is our previously published fit ([1]).

i) events whose $m_{1,2}$ is obtained by combining nearly equal values of E^{had}

$$\{|(E_1^{had} - E_2^{had})/(E_1^{had} + E_2^{had})| \leqslant 15\%\};$$

ii) events whose $m_{1,2}$ is obtained by combining very different values of E^{had}

$$\{|(E_1^{had} - E_2^{had})/(E_1^{had} + E_2^{had})| \geqslant 35\%\}.$$

The values of $\langle n_{ch} \rangle$ *vs.* $m_{1,2}$ for these two samples are shown in fig. 7a) and b).

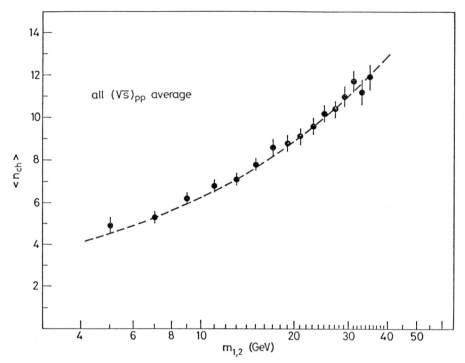

Fig. 5. – Mean charged-particle multiplicity $\langle n_{\text{ch}} \rangle$ *vs.* $m_{1,2}$ averaged over all $(\sqrt{s})_{\text{pp}}$ values. The dashed line is our previously published fit [1].

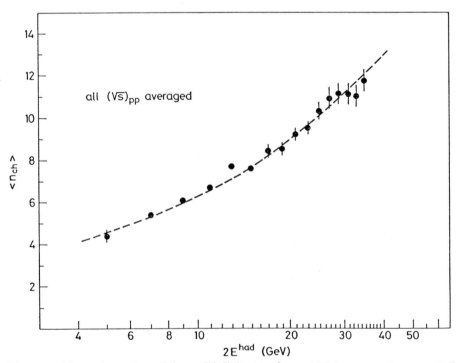

Fig. 6. – Mean charged-particle multiplicity $\langle n_{\text{ch}} \rangle$ *vs.* $2E^{\text{had}}$ averaged over all $(\sqrt{s})_{\text{pp}}$ values. The dashed line is our previously published fit [1].

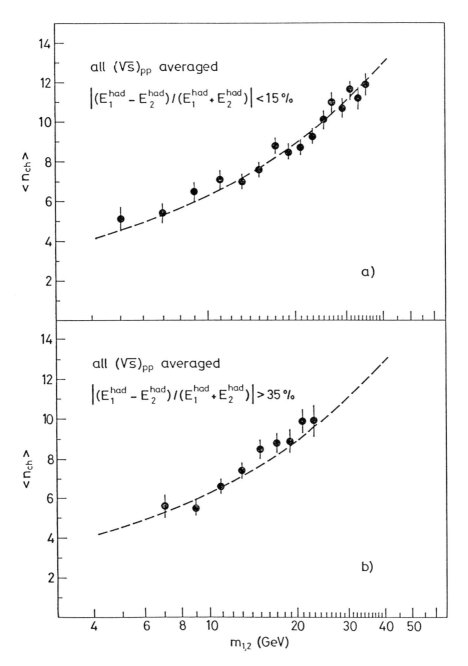

Fig. 7. – Mean charged-particle multiplicity $\langle n_{\mathrm{ch}} \rangle$ vs. $m_{1,2}$ averaged over all $(\sqrt{s})_{pp}$ values, but with the selection a) $|(E_1^{\mathrm{had}} - E_2^{\mathrm{had}})/(E_1^{\mathrm{had}} + E_2^{\mathrm{had}})| \leqslant 15\%$, b) $|(E_1^{\mathrm{had}} - E_2^{\mathrm{had}})/(E_1^{\mathrm{had}} + E_2^{\mathrm{had}})| \geqslant 35\%$. The dashed line is our previously published fit [1].

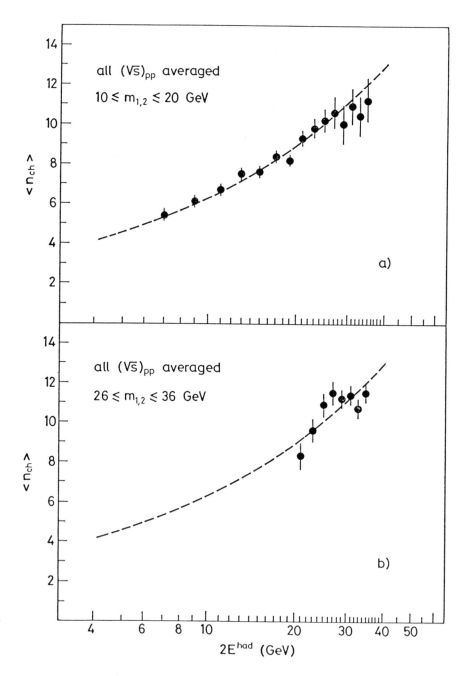

Fig. 8. – Mean charged-particle multiplicity $\langle n_{ch} \rangle$ vs. $2E^{had}$ averaged over all $(\sqrt{s})_p$ values for events with two protons, with the selection a) $10 \leqslant m_{1,2} \leqslant 20$ GeV, b) $26 \leqslant \leqslant m_{1,2} \leqslant 36$ GeV. The dashed line is our previously published fit [1].

For the sake of simplicity, only the values of $\langle n_{\text{ch}} \rangle$ averaged over the different $(\sqrt{s})_{\text{pp}}$ are reported. No significant differences appear in the comparison of these two samples.

A second test, complementary to the first one, has also been done, *i.e.* the study of the influence, on the values of $\langle n_{\text{ch}} \rangle$ *vs.* $2E^{\text{had}}$, of the quantity $m_{1,2}$. Two samples of events with $m_{1,2}$ in the intervals $10 \leqslant m_{1,2} \leqslant 20$ GeV and $26 \leqslant \leqslant m_{1,2} \leqslant 36$ GeV were selected; fig. 8a) and b) show $\langle n_{\text{ch}} \rangle$, averaged over the three values of $(\sqrt{s})_{\text{pp}}$, for these two samples. Again no significant differences appear. This study shows that there is no correlation either in terms of $m_{1,2}$ or in terms of $2E^{\text{had}}$.

3. – Conclusion.

The results of the present experiment show that the average charged-particle multiplicities $\langle n_{\text{ch}} \rangle$, measured in terms of $m_{1,2}$ (fig. 5) and of $2E^{\text{had}}$ (fig. 6), agree with each other and with e^+e^- data. The $(\sqrt{s})_{\text{pp}}$ independence is valid in both cases.

● RIASSUNTO

Usando interazioni pp a tre diverse energie nel centro di massa, $(\sqrt{s})_{\text{pp}} = 30, 44, 62$ GeV, si mostra che la molteplicità media delle particelle cariche $\langle n_{\text{ch}} \rangle$ in funzione della massa invariante del sistema adronico $m_{1,2}$ ha lo stesso andamento che in funzione di $2E^{\text{had}}$. Si mostra inoltre che, in entrambi i casi, $\langle n_{\text{ch}} \rangle$ è pressoché indipendente da $(\sqrt{s})_{\text{pp}}$ e in buon accordo con la molteplicità media delle particelle cariche misurata nel processo di annichilazione (e^+e^-).

Подробное исследование зависимости $\langle n_{\text{ch}} \rangle$ от E^{had} и $m_{1,2}$ при различных $(\sqrt{s})_{\text{pp}}$ в (pp) взаимодействиях.

Резюме (*). — Используя (pp) взаимодействия при трех различных энергиях в системе центра масс, $(\sqrt{s})_{\text{pp}} = 30, 44, 62$ ГэВ, показывается что зависимость средней множественности заряженных частиц $\langle n_{\text{ch}} \rangle$ от инвариантной массы адронной системы $m_{1,2}$ имеет такое же поведение, как и зависимость от $2E^{\text{had}}$. Более того, показывается, что в обоих случаях $\langle n_{\text{pp}} \rangle$ почти не зависит от $(\sqrt{s})_{\text{pp}}$ и $\langle n_{\text{ch}} \rangle$ хорошо согласуется со средней множественностью заряженных частиц, измеренной при (e^+e^-) аннигиляции.

(*) *Переведено редакцией.*

M. BASILE, *et al.*
1 Febbraio 1982
Il Nuovo Cimento
Serie 11, Vol. 67 A, pag. 244-254

CERN
SERVICE D'INFORMATION
SCIENTIFIQUE

Proceedings of the XIX Course of the "Ettore Majorana" International School of Subnuclear Physics, Erice, 1981: The Unity of the Fundamental Interactions (Plenum Press, Inc., New York-London.

WHAT WE CAN LEARN FROM HIGH-ENERGY,

SOFT (pp) INTERACTIONS

M. Basile, G. Bonvicini, G. Cara Romeo, L. Cifarelli,
A. Contin, M. Curatolo, G. D'Ali, B. Esposito, P. Giusti,
T. Massam, R. Nania, F. Palmonari, A. Petrosino, V. Rossi,
G. Sartorelli, M. Spinetti, G. Susinno, G. Valenti,
L. Votano and A. Zichichi

CERN, Geneva, Switzerland
Istituto di Fisica dell'Università di Bologna, Italy
Istituto Nazionale di Fisica Nucleare, Laboratori
    Nazionali di Frascati, Italy
Istituto Nazionale di Fisica Nucleare, Sezione di Bologna
    Italy
Istituto di Fisica dell'Università di Perugia, Italy
Istituto di Fisica dell'Università di Roma, Italy

Presented by A. Zichichi

1. INTRODUCTION

The purpose of this review is to report on a series of similarities between multiparticle hadronic systems produced in (pp) interactions and in (e^+e^-) annihilation[1-12]. The new feature of this work is the fact that, in order to establish these similarities, the basic principle is to evaluate, for each (pp) interaction, the correct energy available for particle production. The old trend was to select high-p_T processes.

One can produce multiparticle hadronic states in basically three ways: i) (hadron-hadron) interactions; ii) (e^+e^-) annihilation; iii) (lepton-hadron) deep inelastic scattering (DIS). As mentioned above, the only way, so far, to correlate the hadronic systems produced in these different ways was to compare the properties, measured in (e^+e^-) and DIS, with high-p_T (hadron-hadron) data. The low-p_T data were considered as examples of typically hadronic phenomena, not suitable for comparison with (e^+e^-) and DIS physics. Our new way of analysing (pp) interactions[1-12] brings into the exciting field of physics the large amount of, so far abandoned, low-p_T (pp) data.

The study of the properties of multihadronic systems produced in low-p_T proton-proton interactions, once the energy for particle production in a (pp) interaction has been correctly calculated, shows that a striking series of analogies can be established between (pp) and (e^+e^-) processes[1-12]. I will not discuss the case of multiparticle production in DIS. The way in which DIS data can be compared with (pp) data will be mentioned at the very end.

The comparison of the multiparticle hadronic systems produced in (pp) interactions and (e^+e^-) annihilation will be discussed in terms of the following five quantities:

  i) the inclusive, single-particle, fractional momentum distribution of the produced particles;
 ii) the inclusive, single particle, transverse momentum distribution of the particles produced;
iii) the average charged particle multiplicity;
 iv) the ratio of "charged" to "total" energy of the multiparticle hadronic systems produced;
  v) the planarity of the multiparticle hadronic systems produced.

2.   EXPERIMENTAL (pp) DATA AND COMPARISON WITH (e^+e^-) DATA

2.1   The Main Points

The experimental data for the (pp) interactions have been taken at the CERN Intersecting Storage Rings (ISR) using a large-volume magnetic field, the so-called Split-Field Magnet (SFM), coupled with a powerful system of multiwire proportional chambers (MWPCs)[13]. It is the nearest system to a bubble-chamber-like instrument at the ISR. The data[1-12] have been collected at three different ISR energies, i.e. $(\sqrt{s})_{pp}$ = 30, 44, and 62 GeV, either using the "minimum" bias mode of running, or triggering on fast particles to enrich the sample of events with "leading" protons.

The comparison of our (pp) data with (e^+e^-) is made using (e^+e^-) data from PETRA, from SPEAR, and from ADONE[14-21].

The key point of our analysis can be illustrated as follows: when you have a (pp) collision, the total energy available for particle production is not $(\sqrt{s})_{pp} = 2E_{inc}$, where E_{inc} is the incident energy of each colliding proton. In fact, quite often a large fraction of the primary energy is carried away, by the incoming proton, into the final state. This is the "leading" proton effect; it is there, and not only in the case of a proton interacting with another proton. As we will see later, it is a very general phenomenon, which is present when a hadron interacts, no matter if strongly, electromagnetically or weakly[22],[23]. The "hadron-in-the-final-state" which is coming from the initial state has an energy sharing, with all other particles produced, which is highly priviliged: this "leading" hadron effect must be accounted for correctly in order to compare the properties of the multiparticle hadronic system produced in the interaction. For example, in the (pp) case, if you study the interaction of two protons, each with 31 GeV, the total energy for particle production is not going to be 62 GeV, but a fraction of this. This fraction depends on the energies taken away by the two leading protons in the final state.

The first evidence for analogies between (pp) and (e^+e^-) data[1] was found using (pp) interactions at the nominal ISR energy, $(\sqrt{s})_{pp}$ = 62 GeV. As mentioned above, this "fixed" nominal energy corresponds to a set of "effective" energies, available for particle production, which range from few GeV up to about 60 GeV. In Fig. 1 the relation between the incoming proton beam energy E_{inc} and the effective hadronic energy available for particle production E_{had} is shown, once the experimental cuts have been taken into account. In addition to E_{inc} = 31 GeV, two other cases, E_{inc} = 22 GeV and E_{inc} = = 15 GeV, are also shown. A large "leading" proton effect produces a small effective hadronic energy; conversely, a small "leading" proton effect produces a large effective hadronic energy. It is the variation of the "leading" proton effect that allows different nominal fixed proton beam energies E_{inc} to produce the same effective hadronic energy $2E_{had}$ available for particle production.

The reason why we have collected data at three ISR energies, $(\sqrt{s})_{pp}$ = 30, 44, and 62 GeV, is as follows. It was a crucial point for our study to show that the multiparticle hadronic systems produced in (pp) interactions, with the same values of $2E_{had}$ but with different values of E_{inc}, had the same properties in terms of the five quantities mentioned above, i.e. fractional momentum distribution, transverse momentum properties, average charged particle multiplicity, ratio of charged to total energy, and planarity.

2.2 The Leading Hadron Effect

As anticipated earlier, the "leading" proton effect is an example of a very general phenomenon: the leading hadron effect[22],[23].

Let F(x) be the inclusive single-particle cross-section integrated over p_T:

$$F(x) = \frac{1}{\pi} \int \frac{2E}{\sqrt{s}} \frac{d\sigma^2}{dxdp_T^2} dp_T^2 ,$$

where $x = 2p_L/\sqrt{s}$, with p_L the longitudinal momentum of the hadron under consideration.

M. BASILE ET AL.

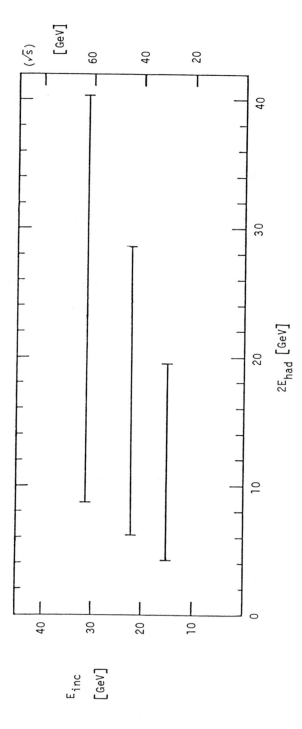

Fig. 1 Ranges of effective hadronic energy available for particle production ($2E_{had}$) for a given ISR incident energy E_{inc}. These ranges depend on the proton x_F range selected (0.35-0.86).

The range $0 \lesssim x < 0.2$ is typical of the so-called "central" production; the range $x > 0.8$ is near to the region where diffractive and quasi-elastic processes start to appear. There is a range of x which is far from the above two well-known physical processes [the so-called "central" production ($x \simeq 0$) and the peripheral production ($x \simeq 1$)]: $0.2 \leq x \leq 0.8$. If we divide this range of x into two parts, $0.2 \leq x \leq 0.4$ and $0.4 < x \leq 0.8$, and if we evaluate the integral of F(x) in these two intervals, the ratio of these two quantities will be a good parameter for expressing, in a quantitative way, the "leading" effect:

$$L = \int_{0.4}^{0.8} F(x)\,dx \left/ \int_{0.2}^{0.4} F(x)\,dx \right. .$$

The quantity L tells us how strong the "leading" effect is in a given process. Figure 2 shows the values of L for different types of particles produced in (pp) interactions at the ISR. The quantity L scales, in the ISR energy range investigated, from $(\sqrt{s})_{pp} = 25$ up to $(\sqrt{s})_{pp} = 62$ GeV. The important point is: the lower the number of propagating quarks the lower the value of L. Notice that the value of L for zero propagating quarks is below 0.5, as in (e^+e^-) annihilation.

A very interesting result is shown in Fig. 3 where the analysis of the ($\bar{p}p$) data, obtained at Fermilab, is given. The values of L show that the leading effect holds true for antiprotons as well as for protons.

Finally, Fig. 4 shows a spectacular result: the leading effect is present in electromagnetic and in weak interactions. In fact, in this figure we report the analysis of Λ^0 production in (ep) and ($\bar{\nu}p$) interactions. The values of L, calculated for the Λ^0, in the electromagnetic production (ep) and in the weak production ($\bar{\nu}p$) processes, are the proof that the Λ^0 is leading.

This synthetic analysis shows that the "leading" proton effect, which is at the basis of our discovery of the similarities between the two ways of producing multiparticle hadronic states, (pp) and (e^+e^-), is a very general phenomenon. It holds true in strong, electromagnetic, and weak interactions, provided a hadron is present in the initial state. The leading effect in the final state is present, even if the initial hadron has transformed itself into another hadron. More transformation implies less leading effect.

We will now discuss the five quantities studied in order to establish the similarities between (pp) and (e^+e^-) processes.

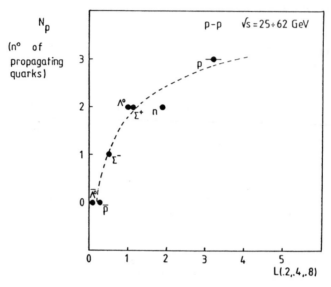

Fig. 2 The leading quantity L(0.2, 0.4, 0.8), for various final-state hadrons in pp collisions at ISR energies (25-62 GeV), is plotted versus the number of propagating quarks from the incoming into the final-state hadrons. The dashed line is obtained by using a parametrization of the single-particle inclusive cross-section, as described in Refs. 22 and 23.

287

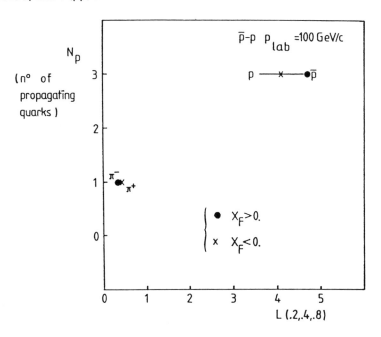

Fig. 3 L(0.2, 0.4, 0.8) for final-state hadrons produced in (p̄-p) collisions at p_{lab} = 100 GeV/c (Refs. 22, 23).

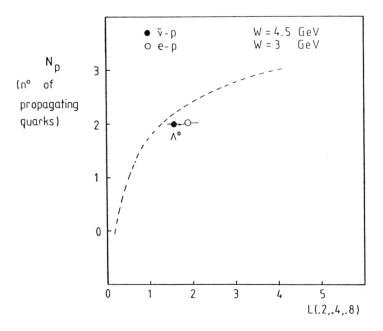

Fig. 4 L(0.2, 0.4, 0.8) for Λ^0 production in (ν̄p) and (ep) reactions. The dashed line is the same as described for Fig. 2 (Refs. 22, 23).

2.3 The Inclusive Fractional Momentum Distribution

The inclusive, single-particle, fractional momentum distributions, $d\sigma/dx_R^*$, of the particles produced in (pp) interactions at the nominal total c.m. energy, $(\sqrt{s})_{pp}$ = 62 GeV[1], are shown in Figs. 5a-5c for different values of the effective hadronic energy E_{had} available for particle production, i.e. once the leading proton effect is correctly accounted for. The fractional momentum x_R^* is evaluated with respect to the effective hadronic energy E_{had}, according to the definition $x_R^* = p/E_{had}$, p being the momentum of the particles. As mentioned above from a sample of (pp) interactions at fixed (pp) energy, $(\sqrt{s})_{pp}$ = = 62 GeV, we have been able to obtain three samples of events at different E_{had} by selecting the leading proton in different energy ranges. The data in Fig. 5a refer to hadronic energies in the range E_{had} = 5-8 GeV; in Fig. 5b the data with E_{had} = 8-11 GeV are reported; in Fig. 5c the data at the highest hadronic energy range, E_{had} = 14-16 GeV, are given. High values of E_{had} means less leading proton effect. At ISR energies it is not easy to identify a "proton" via the Čerenkov effect or via the time-of-flight technique. Our method of identifying a proton in the final state is the simplest: i.e. by its leading property. Experimental measurements at the ISR[24] show that when the proton is more leading, there is less π contamination. For example, if a proton is leading with 80% of its initial momentum, the probability of having a π in the same momentum configuration is $\sim$ 2%. If the proton has 50% of its primary momentum, the π contamination can be as high as $\sim$ 15%. In Fig. 5a the π contamination is expected to be at the 2% level; in Fig. 5c we expect a π contamination of the order of $\sim$ 15%.

In Figs. 5a-5c the (e^+e^-) data[14] at equivalent energies are shown. The agreement is remarkable in Figs. 5a and 5b. The (pp) data in Fig. 5c are systematically below the (e^+e^-) data; this can be understood, as mentioned above, in terms of the unavoidable π contamination present in the (pp) data, in this E_{had} range.

Now we come to the proof that E_{had} is a good variable, independent of the nominal c.m. (pp) energy $(\sqrt{s})_{pp}$: this is shown in Figs. 6a-6c. The first of this set of figures, Fig. 6a, shows the data obtained at three nominal c.m. (pp) energies, namely

$$(\sqrt{s})_{pp} = 30, 44, 62 \text{ GeV} ,$$

but with the same effective hadronic energy range

$$(10 \leq 2E_{had} \leq 16) \text{ GeV} .$$

Irrespective of the primary proton energies, the data on $d\sigma/dx_R^*$ follow the same line. In Fig. 6b, again the same set of (pp) interactions at three values for $(\sqrt{s})_{pp}$ are used, but with a higher hadronic energy range

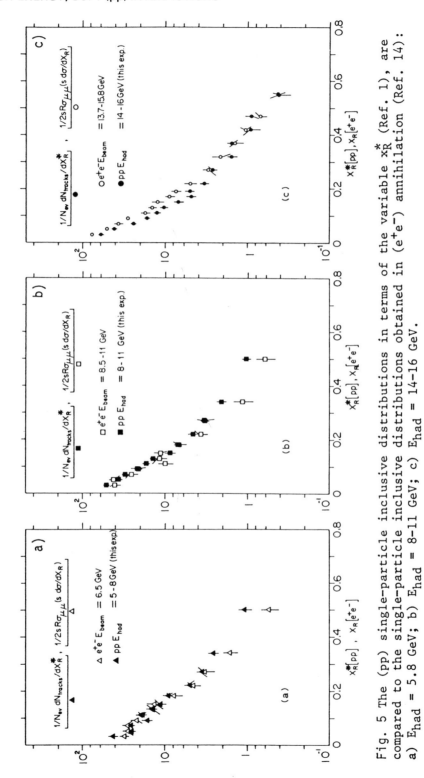

Fig. 5 The (pp) single-particle inclusive distributions in terms of the variable x_R^* (Ref. 1), are compared to the single-particle inclusive distributions obtained in (e^+e^-) annihilation (Ref. 14): a) $E_{had} = 5.8$ GeV; b) $E_{had} = 8-11$ GeV; c) $E_{had} = 14-16$ GeV.

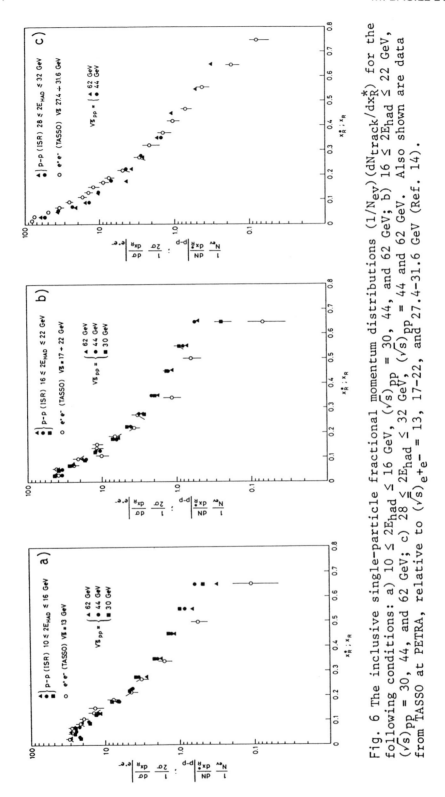

Fig. 6 The inclusive single-particle fractional momentum distributions $(1/N_{ev})(dN_{track}/dx_R^*)$ for the following conditions: a) $10 \le 2E_{had} \le 16$ GeV, $(\sqrt{s})_{pp} = 30$, 44, and 62 GeV; b) $16 \le 2E_{had} \le 22$ GeV, $(\sqrt{s})_{pp} = 30$, 44, and 62 GeV; c) $28 \le 2E_{had} \le 32$ GeV, $(\sqrt{s})_{pp} = 44$ and 62 GeV. Also shown are data from TASSO at PETRA, relative to $(\sqrt{s})_{e^+e^-} = 13$, 17–22, and 27.4–31.6 GeV (Ref. 14).

$$(16 \leq 2E_{had} \leq 22) \text{ GeV .}$$

Finally, in Fig. 6c the lower energy data at $(\sqrt{s})_{pp}$ = 30 GeV cannot be used because the hadronic energy range is too high, i.e.

$$(28 \leq 2E_{had} \leq 32) \text{ GeV .}$$

In the same graphs the (e^+e^-) data[14] at equivalent energies, i.e.

$$(\sqrt{s})_{e^+e^-} = 2E_{had} \text{ ,}$$

are shown for comparison with our data. The agreement is very good.

Having seen that $(d\sigma/dx_R^*)$ does not change with $(\sqrt{s})_{pp}$, we can limit ourselves to the $(\sqrt{s})_{pp}$ = 62 GeV data in order to avoid too many points in a graph. This is done in Fig. 7 in order to show a very interesting fact. If we had had the idea of subtracting the leading proton effect in a (pp) collision earlier, we could have discovered the dramatic rise in $(d\sigma/dx_R^*)$ at low x_R^* values. This dramatic rise in the inclusive fractional momentum distributions has been discovered with (e^+e^-) at PETRA, and it is one of the most important (e^+e^-) results obtained in this energy range. As shown by the SPEAR data at lower (e^+e^-) energy[15], the shape of $(d\sigma/dx_R)$ was flattening at low x_R values.

Now there is a key problem: Who tells us that, going to the equivalent low E_{had} values, $d\sigma/dx_R^*$ follows the same trend as the one observed in the SPEAR energy range? When we first observed the similarity between (pp) and (e^+e^-) data in the quantity $d\sigma/dx_R^*$[1], we had at our disposal ISR data at

$$(\sqrt{s})_{pp} = 62 \text{ GeV.}$$

This is why we could not answer the above question. Using our new ISR data at

$$(\sqrt{s})_{pp} = 30 \text{ GeV}$$

we are able to reach very low hadronic energies[11]. Figures 8a-8c show our data at

$$(3 \leq 2E_{had} \leq 4) \text{ GeV ,}$$
$$(4 < 2E_{had} \leq 6) \text{ GeV ,}$$
$$(6 < 2E_{had} \leq 9) \text{ GeV ,}$$

compared with SPEAR data[15] at equivalent (e^+e^-) energies: $(\sqrt{s})_{e^+e^-} \simeq$ $\simeq 2E_{had}$.

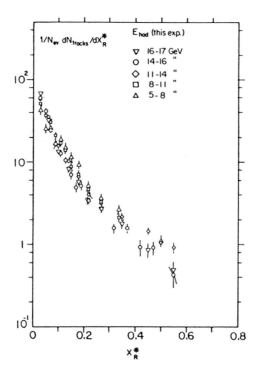

Fig. 7 The inclusive single-particle distributions $(1/N_{ev})(dN_{track}/dx_R^*)$ versus x_R^* for various bands of E_{had}, obtained from (pp) collisions (Ref. 1).

HIGH-ENERGY, SOFT (pp) INTERACTIONS

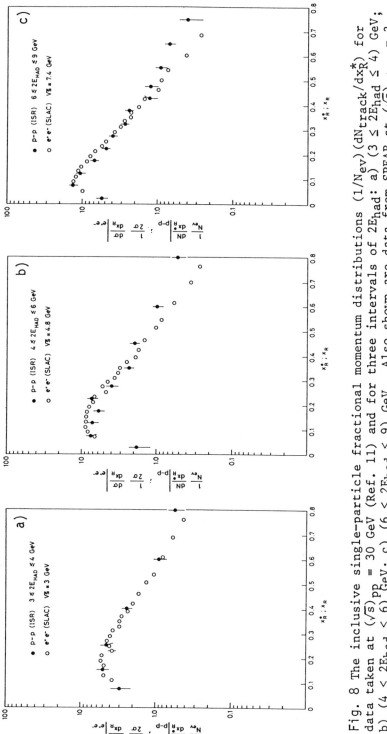

Fig. 8 The inclusive single-particle fractional momentum distributions $(1/N_{ev})(dN_{track}/dx_R^*)$ for data taken at $(\sqrt{s})pp = 30$ GeV (Ref. 11) and for three intervals of $2E_{had}$: a) $(3 \le 2E_{had} \le 4)$ GeV; b) $(4 \le 2E_{had} \le 6)$ GeV; c) $(6 \le 2E_{had} \le 9)$ GeV. Also shown are data from SPEAR at $(\sqrt{s})e^+e^- = 3$, 4.8, and 7.4 GeV (Ref. 15).

The data reported in Figs. 8a-8c show that, also in the very low hadronic energy range, the agreement between our data and (e^+e^-) annihilation is very good. This proves that the dramatic rise in the single-particle inclusive momentum distribution could have been observed in (pp) interactions at the ISR before being discovered in (e^+e^-) annihilation at PETRA.

2.4 The Inclusive Transverse Momentum Distribution

Let me now go to another important property of the multiparticle hadronic systems produced in (pp) interactions; namely, the inclusive transverse momentum distribution of the particles produced in a (pp) interaction where the effective hadronic energy has been correctly calculated. Figure 9 shows our results, obtained with $(\sqrt{s})_{pp} = 30$ GeV data, where a sample of events in the hadronic energy range

$$(11 \leq 2E_{had} \leq 13) \text{ GeV}$$

has been selected.

In Fig. 9 this is compared with (e^+e^-) data at $(\sqrt{s})_{e^+e^-} = 12$ GeV[16]. The agreement is excellent.

A new way of analysing the transverse momentum distribution of the particles produced in (e^+e^-) annihilation has recently been introduced[17]. It is based on the so-called reduced variable $p_T/\langle p_T \rangle$, where p_T is the transverse momentum of the single particle observed and $\langle p_T \rangle$ is the average value of the same quantity for all particles produced.

The results[7] are shown in Fig. 10, where we compare two ranges of hadron energies,

$$(8 \leq 2E_{had} \leq 16) \text{ GeV}$$
$$(24 \leq 2E_{had} \leq 32) \text{ GeV} ,$$

with the (e^+e^-) equivalent energy data obtained by the PLUTO group[17]. The data overlap very well. So even the transverse momentum distributions of the particles produced follow the same trend in (pp) and (e^+e^-).

2.5 The Average Charged Particle Multiplicity

Another important quantity for comparing (pp) and (e^+e^-) is the average number of charged particles produced at a given energy.

The data shown in Fig. 11 refer to the results obtained using three samples of data at $(\sqrt{s})_{pp} = 30$, 44, and 62 GeV[8]. These data have been analysed, on an event-by-event basis, with the method of

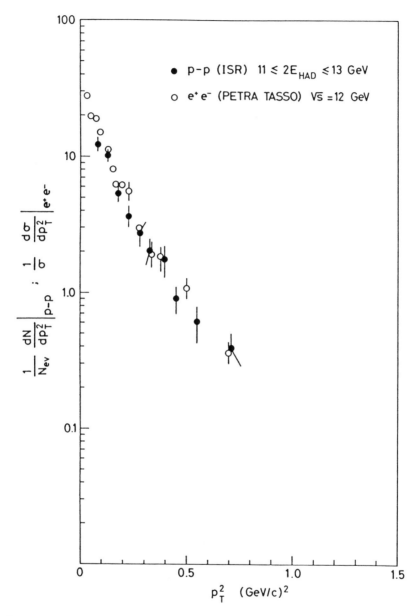

Fig. 9 The inclusive single-particle transverse-momentum distribution $(1/N_{ev})(dN/dp_T^2)$ for $(11 \leq 2E_{had} \leq 13)$ GeV (Ref. 10) compared with $(1/\sigma)/(d\sigma/dp_T^2)$ measured at PETRA (TASSO) at $(\sqrt{s})_{e^+e^-} = 12$ GeV (Ref. 16).

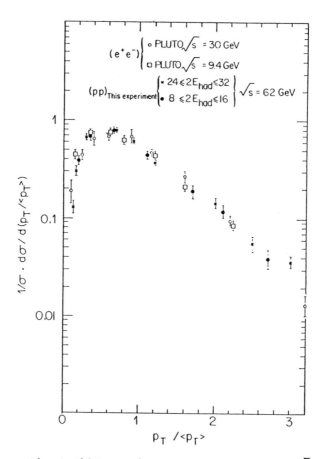

Fig. 10 Renormalized differential cross-section $(1/\sigma)\left[d\sigma/d(p_T/\langle p_T\rangle)\right]$ versus the "reduced" variable $p_T/\langle p_T\rangle$ (Ref. 7). These distributions allow a comparison of the multiparticle systems produced in e^+e^- annihilation (Ref. 17) and in pp interactions in terms of the "reduced" transverse momentum properties.

297

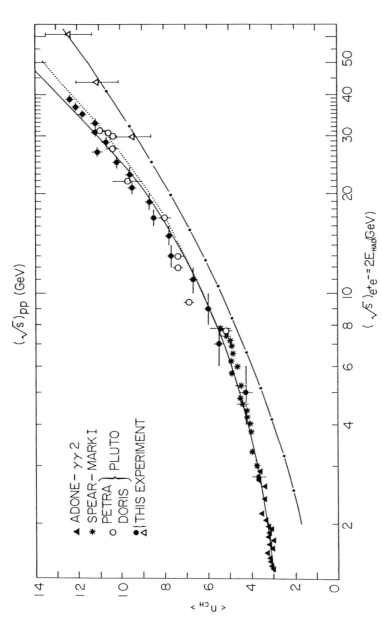

Fig. 11 Mean charged particle multiplicity, averaged over different $(\sqrt{s})_{pp}$, versus $(2E_{had})$ (Ref. 8), compared with (e^+e^-) data (Refs. 15, 18, 19). The contribution of $K_S^0 \rightarrow \pi^+\pi^-$ has been subtracted. Each value is an average over 2 GeV. The quoted errors are statistical only. The systematic uncertainty is less than 8%. The continuous line is the best fit to our data according to the formula $\langle n_{ch} \rangle = a + b \exp\left[c\sqrt{\ln (s/\Lambda^2)}\right]$. The dotted line is the best fit using PLUTO data (Ref. 19). The dashed-dotted line is the (pp) total charged particle multiplicity (Ref. 25). For this curve, the abscissa is $(\sqrt{s})_{pp}$. The open triangular points in this curve are our data using the standard way (i.e. without the subtraction of the leading proton effects), to calculate $\langle n_{ch} \rangle$ at fixed $(\sqrt{s})_{pp}$.

calculating the effective hadronic energy available in the (pp) interaction, E_{had}, after the leading proton effects are subtracted. Once again, the use of different $(\sqrt{s})_{pp}$ to produce the same E_{had} is crucial in order to prove that the quantity E_{had} is a good variable. In the same figure (Fig. 11), the results from (e^+e^-) experiments are also reported[15,18,19]. The agreement between (pp) and (e^+e^-) data is very satisfactory.

It should be remarked that the "standard" way of analysing (pp) data, i.e. without the subtraction of the "leading" proton effect, would produce the dashed-dotted curve plotted in Fig. 11[25], where the (e^+e^-) data clearly stay above the standard (pp) ones. An important cross-check is given by the three points (open triangles) plotted in Fig. 11. These points are the results of our analysis at $(\sqrt{s})_{pp}$ = 30, 44, and 62 GeV, once we proceed in the "standard" way.

So, even with an important quantity such as the number of charged particles produced on the average in (e^+e^-) annihilation and in (pp) interactions at equivalent energies, we find an excellent agreement, once the energy available for particle production in a (pp) interaction is correctly calculated.

2.6   The Ratio of Charged to Total Energy

In a (pp) interaction, studied in the standard way, the total energy is $(\sqrt{s})_{pp}$. If we now try to calculate the energy carried away by the charged particles, and make the ratio

$$\frac{\langle \text{Energy carried by the charged particles} \rangle}{\text{Total energy}} ,$$

we find no agreement between this value and the ratio

$$\frac{\langle E_{charged} \rangle}{E_{total}}$$

measured in (e^+e^-) annihilation.

The method of removing the leading proton effects makes it possible, in (pp) interactions, to calculate on an event-by-event basis the effective hadronic energy E_{had}, i.e. the total hadronic energy which is really useful for that given (pp) interaction. On the other hand, the energy carried by the charged particles, $E_{charged}$, is directly measured. From these two quantities, measured for each event, the average value $\langle E_{charged} \rangle$ can be calculated, and the ratio $\langle E_{charged} \rangle / E_{had} = \alpha_{pp}$ can be compared with $\langle E_{charged} \rangle / E_{total}$ measured in (e^+e^-) annihilation.

The (pp) results[5] are shown in Fig. 12 together with the (e^+e^-) data[15,20]. Once again we use three sets of data at $(\sqrt{s})_{pp}$ = 30, 44, and 62 GeV to produce events with equal E_{had} values.

The results of Fig. 12 show that the average fractional energy available for charged particle production in a (pp) interaction is, within the experimental uncertainty, the same as for the (e^+e^-) annihilation.

2.7 The Event Planarity

The problem under investigation is now the following: Do the events produced in a (pp) interaction show a planarity structure? To answer this question we proceed[4] in the same way as our colleagues working on (e^+e^-), adopting the same terminology and the same symbols. Let me remind you about a few basic points.

For each event we

i) construct a quantity (see Fig. 13)

$$M_{\alpha\beta} = \sum_{j=1}^{N} P_{j\alpha} P_{j\beta} \qquad (\alpha,\beta = 1,2) \; ,$$

where N = number of particles in the event;

ii) determine the eigenvectors $\vec{n}_1$, $\vec{n}_2$, and eigenvalues Λ_1, Λ_2 ($\Lambda_1 < \Lambda_2$);

$\vec{n}_1$ is the direction in which the sum of the square of the momentum components is minimized;

$\vec{n}_2$ together with the (pp) line of flight, defines the event plane.

The next step is to define the average p_T^2 "in" and "out" of the event plane, according to

$$\langle p_T^2 \rangle_{out} = \frac{\Lambda_1}{N} = \frac{1}{N} \sum_{j=1}^{N} (\vec{p}_j \cdot \vec{n}_1)^2 \; ,$$

$$\langle p_T^2 \rangle_{in} = \frac{\Lambda_2}{N} = \frac{1}{N} \sum_{j=1}^{N} (\vec{p}_j \cdot \vec{n}_2)^2 \; .$$

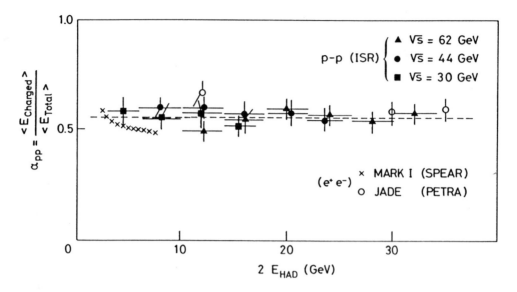

Fig. 12 The charged-to-total energy ratio obtained in pp collisions (Ref. 5), α_{pp}, plotted versus $2E_{had}$ and compared with e^+e^-
data obtained at SPEAR (Ref. 15) and PETRA (Ref. 20).

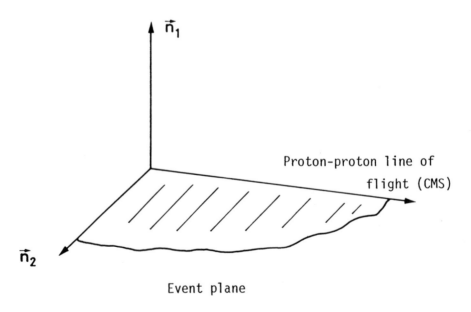

Fig. 13 Event plane definition.

Figure 14 shows $\langle p_T^2 \rangle_{in}$ and $\langle p_T^2 \rangle_{out}$ distributions in the low-energy range $(10 \leq 2E_{had} \leq 16)$ GeV[4]. The equivalent (e^+e^-) energies are $\left[13 \leq (\sqrt{s})_{e^+e^-} \leq 17 \right]$ GeV. The agreement between (pp) and (e^+e^-) data[21] is very good.

Figure 15 shows the results[4] at higher energy, i.e. $(28 \leq 2E_{had} \leq 34)$ GeV. Notice that the $\langle p_T^2 \rangle_{out}$ distribution follows the Monte Carlo prediction. The deviation of the $\langle p_T^2 \rangle_{in}$ distribution from Monte Carlo predictions is the proof that the events produced in (pp) interactions do show a planarity effect. Notice that the (e^+e^-) data[21] stay above the (pp) data for the "in" and the "out" distributions. If this effect is going to be confirmed by more accurate (pp) and (e^+e^-) data, the interpretation could be in terms of "heavy" quarks being produced in (e^+e^-) more easily than in (pp) interactions. In fact, once the energy is over the threshold, the production of a quark depends on its charge squared in (e^+e^-) annihilation. However, the phenomenological trend in (pp) interactions, also expected from QCD, is in terms of the inverse mass squared of the quark to be produced. This implies that in (e^+e^-) the production of heavy quarks competes well with the production of light quarks, but in (pp) interactions the light quark production dominates the "heavy" quark one. The results at different E_{had} values, reported in Fig. 14 and 15, are all from the same (pp) nominal energy $(\sqrt{s})_{pp} = 62$ GeV.

We now come to a very interesting point; namely, the $(\sqrt{s})_{pp}$ independence of the planarity effect. You can see in Fig. 16 and 17 that the data obtained from different values of $(\sqrt{s})_{pp}$, i.e. $(\sqrt{s})_{pp} = 30, 44, 62$ GeV, overlap each other, and show the same planarity structure.

2.8 Conclusions

Let me summarize the most important points.

The hadronic production in (e^+e^-) annihilation takes place in such a way that no hadron has a privileged energy sharing. The inclusive fractional momentum distribution of the particles produced in (e^+e^-) annihilation follows the same shape as for hadronic systems produced in (pp) interactions[1]. This similarity can be established only if the "leading" proton effects are removed[1]. Thus the "leading" effect plays an important role in the understanding of the similarities and therefore of the connection between these two different ways of producing multiparticle hadronic systems[1-12]. The leading effect is a very general phenomenon[22,23]. It shows up whenever there is a hadron in the initial state of an interaction -- no matter whether the interaction is strong, electromagnetic, or weak -- no matter whether or not the initial hadron remains exactly as it is. For example, the leading effect is present if a proton becomes a neutron, or a Λ_s^0, or even a Λ_c^+. Even if the quantum numbers of the hadron in the initial state are not fully carried out in the final-state hadron,

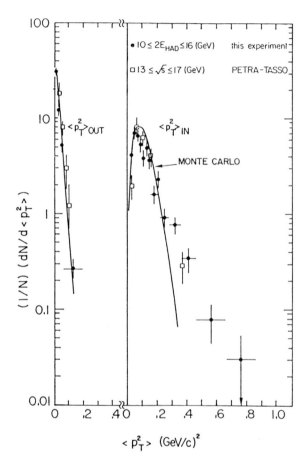

Fig. 14 The average transverse-momentum squared distributions for the "out" and the "in" cases, in the "low-energy" range (10 $\leq$ 2E$_{had}$ $\leq$ $\leq$ 16) GeV (Ref. 4), compared with (e$^+$e$^-$) data at [13 $\leq$ ($\sqrt{s}$)$_{e^+e^-}$ $\leq$ $\leq$ 17] GeV (Ref. 21). The full lines are the results of a p$_T$-limited phase-space Monte Carlo.

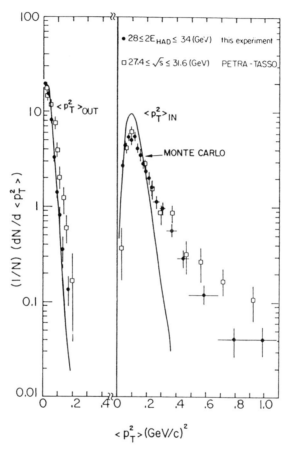

Fig. 15 The average transverse-momentum squared distributions for the "out" and the "in" cases, in the "high-energy" range ($28 \leq 2E_{had} \leq \leq 34$) GeV (Ref. 4), compared with (e^+e^-) data at $[27.4 \leq (\sqrt{s})_{e^+e^-} \leq 31.6]$ GeV (Ref. 21). The full lines are the results of a p_T-limited phase-space Monte Carlo.

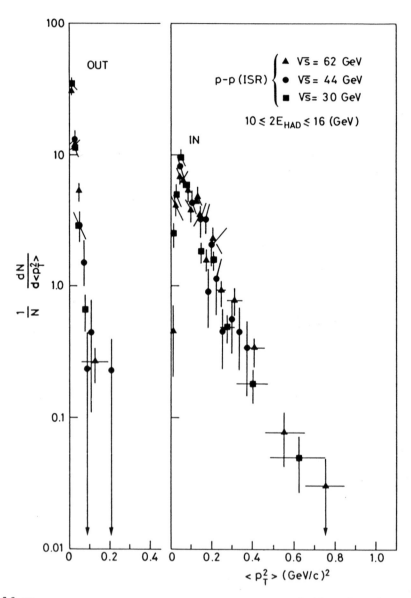

Fig. 16 The average transverse-momentum squared distributions for the "out" and the "in" cases. Data relative to the effective hadronic energy interval $(10 \leq 2E_{had} \leq 16)$ GeV obtained from (pp) collisions at $(\sqrt{s})_{pp} = 30, 44,$ and 62 GeV are shown.

HIGH-ENERGY, SOFT (pp) INTERACTIONS   719

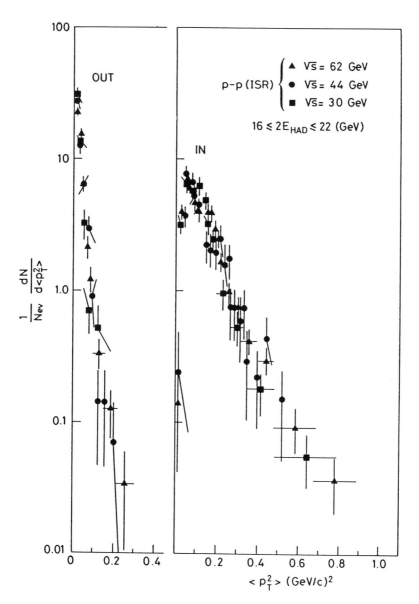

Fig. 17 The average transverse-momentum squared distributions for the "out" and the "in" cases. Data relative to the effective hadronic energy interval (16 ≤ 2E$_{had}$ ≤ 22) GeV obtained from (pp) collisions at $(\sqrt{s})_{pp}$ = 30, 44, and 62 GeV are shown.

the leading effect is present. It is the quantum number flow, be it colour, flavour, J^{PC}, etc., which gives to the initial-state hadron a highly privileged energy sharing, when compared with the particles produced.

It should therefore be clear that all reactions where an initial-state hadron is present should be analysed by taking into account the "leading" phenomenon. For example, the final-state hadronic system produced in DIS should also be analysed in such a way as to identify the leading hadron and to subtract its effects from the global inter-action. In the present study we could not report on the comparison between (pp) and deep inelastic lepton-hadron scattering (DIS) because the DIS data have so far never been analysed by subtracting the pre-ence of a leading hadron in the final state.

The use of the leading hadron effect in (pp) interactions has brought the vast amount of low-p_T physics into the very interesting domain of comparison with (e^+e^-) physics[1-12]. Let me emphasize, once again, that the (pp) data presented to you are obtained with the sim-plest possible logic: in a proton-proton collision try to do nothing, as would be the case with a bubble chamber, and look at the (pp) event. This means: study the lowest possible p_T interactions. This has been done in terms of the five quantities investigated: i) the inclusive fractional momentum distribution; ii) the inclusive transverse momen-tum distribution; iii) the average charged particle multiplicity; iv) the ratio of charged-to-total energy; v) the planarity. The low-p_T physics appears to be as interesting as the high-p_T physics. So, the old belief that only high-p_T (pp) phenomena could be the basis for comparisons with (e^+e^-) seems to be over. In order to understand the way in which the multiparticle hadronic systems are produced in strong, electromagnetic, and weak interactions, all p_T physics needs to be investigated.

REFERENCES

1.   M. Basile, G. Cara Romeo, L. Cifarelli, A. Contin, G. D'Ali,
       P. Di Cesare, B. Esposito, P. Giusti, T. Massam,
       F. Palmonari, G. Sartorelli, G. Valenti and A. Zichichi,
       Phys. Lett. 92B 367 (1980).
2.   M. Basile, G. Cara Romeo, L. Cifarelli, A. Contin, G. D'Ali,
       P. Di Cesare, B. Esposito, P. Giusti, T. Massam, R. Nania,
       F. Palmonari, G. Sartorelli, G. Valenti and A. Zichichi,
       Nuovo Cimento 58A 193 (1980).
3.   M. Basile, G. Cara Romeo, L. Cifarelli, A. Contin, G. D'Ali,
       P. Di Cesare, B. Esposito, P. Giusti, T. Massam, R. Nania,
       F. Palmonari, G. Sartorelli, G. Valenti and A. Zichichi,
       Phys. Lett. 95B 311 (1980).

4. M. Basile, G. Cara Romeo, L. Cifarelli, A. Contin. G. D'Ali,
    P. Di Cesare, B. Esposito, P. Giusti, T. Massam, R. Nania,
    F. Palmonari, G. Sartorelli, G. Valenti and A. Zichichi,
    Nuovo Cimento Lett. 29 491 (1980).
5. M. Basile, G. Cara Romeo, L. Cifarelli, A. Contin, G. D'Ali,
    P. Di Cesare, B. Esposito, P. Giusti, T. Massam, R. Nania,
    F. Palmonari, G. Sartorelli, M. Spinetti, G. Susinno,
    G. Valenti and A. Zichichi, Phys. Lett. 99B 247 (1981).
6. M. Basile, G. Cara Romeo, L. Cifarelli, A. Contin, G. D'Ali,
    P. Di Cesare, B. Esposito, P. Giusti, T. Massam, R. Nania,
    F. Palmonari, G. Sartorelli, M. Spinetti, G. Susinno,
    G. Valenti and A. Zichichi, Nuovo Cimento Lett. 30 389
    (1981).
7. M. Basile, G. Cara Romeo, L. Cifarelli, A. Contin, G. D'Ali,
    P. Di Cesare, B. Esposito, P. Giusti, T. Massam, R. Nania,
    F. Palmonari, G. Sartorelli, M. Spinetti, G. Susinno,
    G. Valenti, L. Votano and A. Zichichi, Nuovo Cimento Lett.
    31 273 (1981).
8. M. Basile, G. Cara Romeo, L. Cifarelli, A. Contin, G. D'Ali,
    P. Di Cesare, B. Esposito, P. Giusti, T. Massam, R. Nania,
    F. Palmonari, V. Rossi, G. Sartorelli, M. Spinetti,
    G. Susinno, G. Valenti, L. Votano and A. Zichichi, Nuovo
    Cimento 65A 400 (1981).
9. M. Basile, G. Cara Romeo, L. Cifarelli, A. Contin, G. D'Ali,
    P. Di Cesare, B. Esposito, P. Giusti, T. Massam, R. Nania,
    F. Palmonari, V. Rossi, G. Sartorelli, M. Spinetti,
    G. Susinno, G. Valenti, L. Votano and A. Zichichi, Nuovo
    Cimento 65A 414 (1981).
10. M. Basile, G. Cara Romeo, L. Cifarelli, A. Contin, G. D'Ali,
    P. Di Cesare, B. Esposito, P. Giusti, T. Massam, R. Nania,
    F. Palmonari, V. Rossi, G. Sartorelli, M. Spinetti,
    G. Susinno, G. Valenti, L. Votano and A. Zichichi, Nuovo
    Cimento Lett. 32 210 (1981).
11. M. Basile, G. Cara Romeo, L. Cifarelli, A. Contin, G. D'Ali,
    P. Di Cesare, B. Esposito, P. Giusti, T. Massam, R. Nania,
    F. Palmonari, V. Rossi, F. Rohrback, G. Sartorelli,
    M. Spinetti, G. Susinno, G. Valenti, L. Votano and
    A. Zichichi, Nuovo Cimento 67A 53 (1982).
12. M. Basile, G. Bonvicini, G. Cara Romeo, L. Cifarelli, A. Contin,
    M. Curatolo, G. D'Ali, P. Di Cesare, B. Esposito, P. Giusti,
    T. Massam, R. Nania, F. Palmonari, A. Petrosino, V. Rossi,
    G. Sartorelli, M. Spinetti, G. Susinno, G. Valenti, L. Votano
    and A. Zichichi, preprint CERN-EP/81-146 (1981), submitted
    to Nuovo Cimento.
13. R. Bouclier, R.C.A. Brown, E. Chesi, L. Dumps, H.G. Fischer,
    P.G. Innocenti, G. Maurin, A. Minten, L. Nauman, F. Piuz and
    O. Ullaland, Nucl. Instrum. Methods 125 19 (1975).
14. TASSO Collaboration (R. Brandelik et al.), Phys. Lett. 89B 418
    (1980).

15.   J.L. Siegrist, Ph.D. Thesis, SLAC Report No. 225 (1979).
16.   G. Wolf, DESY Report 80/85 (1980).
17.   PLUTO Collaboration (Ch. Berger et al.), DESY Report 80/111
      (1980).
18.   C. Bacci, G. de Zorzi, G. Penso, B. Stella, D. Bollini,
      R. Baldini Celio, G. Battistoni, G. Capon, R. Del Fabbro,
      E. Iarocci, M.M. Massai, S. Moriggi, G.P. Murtas, M. Spinetti
      and L. Trasatti, Phys. Lett. 86B 234 (1979).
19.   PLUTO Collaboration (Ch. Berger et al.), Phys. Lett. 95B 313
      (1980).
20.   JADE Collaboration (W. Bartel et al.) DESY Report 80/46 (1980).
21.   TASSO Collaboration (R. Brandelik et al.), Phys. Lett. 86B 243
      (1979).
22.   M. Basile, G. Cara Romeo, L. Cifarelli, A. Contin, G. D'Ali,
      P. Di Cesare, B. Esposito, P. Giusti, T. Massam, R. Nania,
      F. Palmonari, V. Rossi, G. Sartorelli, M. Spinetti,
      G. Susinno, G. Valenti, L. Votano and A. Zichichi, Nuovo
      Cimento 66A 129 (1981).
23.   M. Basile, G. Cara Romeo, L. Cifarelli, A. Contin, G. D'Ali,
      P. Di Cesare, B. Esposito, P. Giusti, T. Massam, R. Nania,
      F. Palmonari, V. Rossi, G. Sartorelli, M. Spinetti,
      G. Susinno, G. Valenti, L. Votano and A. Zichichi, Nuovo
      Cimento Lett. 32 321 (1981).
24.   J.W. Chapman, J.W. Cooper, N. Green, A.A. Seidl,
      J.C. Vender Velde, C.M. Bromberg, D. Cohen, T. Ferbel and
      P. Slattery, Phys. Rev. Lett. 32 257 (1974).
      P. Capiluppi, G. Giacomelli, A.M. Rossi, G. Vannini and
      A. Bussière, Nucl. Phys. B70 1 (1974).
25.   W. Thomé, K. Eggert, K. Giboni, H. Lisken, P. Darriulat,
      P. Dittman, M. Holder, K.T. McDonald, H. Albrecht, T. Modis,
      K. Tittel, H. Preissner, P. Allen, J. Derado, V. Eckardt,
      H.-J. Gebauer, R. Meinke, P. Seyboth and S. Uhlig, Nucl.
      Phys. B129 365 (1977).
      A. Albini, P. Capiluppi, G. Giacomelli and A.M. Rossi, Nuovo
      Cimento 32A 101 (1976).

DISCUSSIONS

CHAIRMAN : A. Zichichi

Scientific Secretaries: G. D'Ali, G. Sartorelli

DISCUSSION 1

- SEIBERG:

You described a method for analysing hadronic processes, defining an "effective energy" in terms of which proton-proton interactions behave like e^+e^- processes. What is the maximum effective energy you can obtain in this way and what are the applications of this energy?

- ZICHICHI:

The range of hadronic effective energies available varies according to a) the x_F cut one applies to the leading proton; b) the total energy ($\sqrt{s}$) available in the pp collision. For example, one can select the leading proton in the range

$$0.35 \leq x_F \leq 0.86 \ .$$

In this case, the effective hadronic energy available for various values of $\sqrt{s}$ in the proton-proton centre-of-mass system is as follows (see Fig. 1 of the lecture):

$$\sqrt{s} = 30 \text{ GeV} \qquad 4.2\text{--}19 \text{ GeV} \ ,$$
$$\sqrt{s} = 44 \text{ GeV} \qquad 6.2\text{--}28 \text{ GeV} \ ,$$
$$\sqrt{s} = 62 \text{ GeV} \qquad 8.7\text{--}40 \text{ GeV} \ .$$

When you have two protons colliding with energy E_1 and E_2, $E_1 = E_2 = E_{inc}$, the total c.m. energy available in this collision is generally taken to be
$$\sqrt{s} = E_1 + E_2 = 2E_{inc}.$$

However, we say that this is not the right energy to use in analysing the process. After the collision, the two incoming protons keep, on the average, a large fraction of the available energy and play a privileged role in the energy-momentum sharing among the particles in the final state. Our statement is that you have to subtract, from the total energy, the energy carried by the two leading protons in order to obtain the effective energy available for particle production. We call it E^{had}, with $E^{had} = E_1^{had} + E_2^{had}$. (The reason for this splitting into two terms will soon be clear).

There are many ways of identifying this energy experimentally. The best one is obtained when you identify both protons in the final state. In this case, and with the definitions

four momentum of incoming proton 1 $\equiv p_1^{inc}$

four momentum of incoming proton 2 $\equiv p_2^{inc}$

four momentum of leading proton 1 $\equiv p_1^{lead}$

four momentum of leading proton 2 $\equiv p_2^{lead}$

$$p_1^{had} = p_1^{inc} - p_1^{lead} \quad \text{and} \quad p_2^{had} = p_2^{inc} - p_2^{lead}$$

you can construct the relativistic invariant quantity

$$Q_{1,2}^2 = \left(p_1^{had} + p_2^{had}\right)^2 \quad \text{and} \quad E_{1,2}^{had} = \sqrt{Q_{1,2}^2} \, .$$

However, to a good approximation the analysis can be done quite satisfactorily in terms of the simpler energy variable,

$$\begin{cases} E_1^{had} = E_1^{inc} - E_1^{lead} \\ \\ E_2^{had} = E_2^{inc} - E_2^{lead} \end{cases} \quad \text{and} \quad E_{1,2}^{had} \cong E_1^{had} + E_2^{had} \, .$$

Let me come now to the important point, namely the limits of this analysis. In proton-proton collisions at ISR energies you do not have Čerenkov counters which can discriminate 25 GeV/c protons, nor any other system such as time of flight, so that you can only rely on nature. Nature has been kind enough to give us a very powerful tool in terms of the behaviour of the proton in this kind of inter-action.

The ratio p/π in pp collisions at the ISR behaves exponentially (see Fig. 18) as a function of x_F, and scales with $\sqrt{s}$ and p_T. So, if you take, for example, particles with $x_F = 0.5$ you will find a sample of $\sim 85\%$ protons with $\sim 15\%$ π contamination.

So the proton identification can be done by finding the fastest positive particle in an event and applying to it a lower x_F cut. We

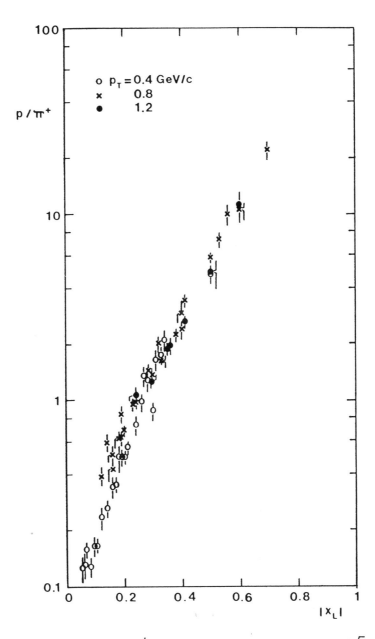

Fig. 18 The inclusive p/π^+ ratio measured at the ISR ($\sqrt{s}$ = 23, 31, 45, 53 GeV).

choose $x_F > 0.4$ as a safe cut. Notice that where the proton has $x_F \simeq 0.8$ you have only a few percent of π contamination, and only 20% of the initial energy is used for particle production.

In this case a high-energy pp collision at, say, 62 GeV gives an effective energy of only 12 GeV, and this is the e^+e^- energy you have to take for a comparison.

DISCUSSION 2   (Scientific Secretaries: G. D'Ali, G. Sartorelli)

- *HERTEN:*

Could you describe the event selection. What cuts do you use in the analysis? and how are the leading protons separated from all other hadrons?

- *ZICHICHI:*

A first sample of data has been collected using the so-called minimum bias trigger -- where you require nothing but a real proton-proton interaction (two tracks in the apparatus) -- in such a way as to minimize beam-gas background and interactions with the beam-pipe wall. In this sample the analysis was performed looking at the jet on one side only and making assumptions on the other side.

There was in fact only one proton per event identified in most of the triggered events. The idea was that this could be an approximation that would not have destroyed the meaning of the method and the results.

A second run has been done, selecting events rich in leading protons. It has been checked that this trigger did not bias the event; in other words, one has only to make sure one does not lose, for acceptance reasons, the leading proton, without which one cannot define the right energy available for particle production.

- *HERTEN:*

Is the energy of the leading proton always so high that it is possible to distinguish it from all the other hadrons produced?

- *ZICHICHI:*

This is an important question. As I have already explained, we limit our analysis to the x_F region

$$0.35 \leq x_F \leq 0.86$$

and this restricts our effective hadronic energy range. To go below this lower limit you would need some kind of particle identification system (Čerenkov counters or a TOF system). We are seriously thinking of the possibility of introducing such a system and of extending our study to the low x_F (high E_{had}) region.

I am personally convinced that pp machines are really very good: it is very difficult to work with them, but, once you disentangle the right physics you want to do, they compete very well with other sources of particles such as e^+e^- machines.

DISCUSSION 3   (Scientific Secretaries: G. D'Ali, G. Sartorelli)

- *KARLINER:*

As you mentioned in your talk, in pp scattering (unlike in e^+e^- scattering) gluon-gluon scattering is almost unavoidable. Is there any chance of seeing the signature of the glueball pole in that kind of scattering?

- *ZICHICHI:*

If we find this effect, I will invite you to participate in a future "Ettore Majorana" School, free of admission fee! It would be extremely difficult to isolate this kind of effect. We have been looking at all possible sources of differences which could show up in charged particles produced -- higher transverse momenta, charge-correlations etc. -- but, up to now, we have not been able to find any.

- *KARLINER:*

As you know, some theorists have estimated the glueball mass based on J/ψ decay. Has anyone seriously tried to see what its effects would be in pp scattering?

- *ZICHICHI:*

I think that the trouble with the glueball mass is that there has been no adequate theoretical investigation of the problem. However, it is not easy to work with glueballs. Sam Lindenbaum worked a lot on this problem and he claims he has evidence for glueballs.

DISCUSSION 4   (Scientific Secretaries: G. D'Ali, G. Sartorelli)

- *WIGNER:*

You reported on a interesting similarity in jets and particle production between pp and e^+e^- collisions, when processes at proper energies are compared. Are there similar results for ep collision products?

- *ZICHICHI:*

This is a field which the previous speaker is working on, and the interaction being studied is not ep but μp. The EMC Collaboration has been investigating the production of hadronic systems (jets) in the collision between space-like virtual photons (emitted by the muon) and the proton. The recoiling proton should be "leading" in the γp c.m. system.

314

728   M. BASILE ET AL.

My statement is based on the analysis we performed on data for the process

$$\bar{\nu}_\mu + p \rightarrow \Lambda^0 + X .$$

In this case we found that the Λ^0 is "leading" in the sense defined previously. The reaction is

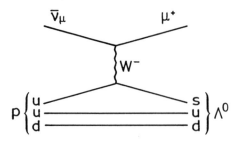

The W^- interacts with the u quark, converting it into an s quark which, with the remaining u and d quarks of the incoming proton, forms the Λ^0. The latter then goes on in the same direction as the incoming proton with a significant fraction of the incoming energy. In terms of our variable L the value of L_{Λ^0} in this case is

$$L_{\Lambda^0} \simeq 2 .$$

in agreement with what I showed you for the values of L for leading particles. This proves that the Λ^0 produced in weak interactions is leading.

So when the EMC Collaboration will analyse their data, identifying the hadrons, they will find a "leading" baryon, and if they analyse the final state in terms of the effective energy we have introduced, I think they will discover the same structures we claim are common to e^+e^- and pp final states.

  - *GABATHULER:*

The leading proton effect in μp collisions corresponds to the target proton in the backward hemisphere in the c.m. system. In the EMC experiment I reported, the detector collected only events in the forward hemisphere. At present we have added a vertex detector consisting of magnet and streamer chamber to study the backward hemisphere, i.e. the leading proton.

  - *YAMAMOTO:*

The data you showed seem to suggest that leading protons have nothing to do with jet formation. Is it true?

- *ZICHICHI:*

To some extent we can say that leading protons have nothing to do with the formation of jets. We have been looking for some correlations and we are still studying the problem since it is an interesting one, but so far we can only say that there are no correlations but kinematics.

- *YAMAMOTO:*

What is the connection with the Drell-Yan process which suggests that there are forward and backward jets and two away-side jets?

- *ZICHICHI:*

The Drell-Yan process itself is an annihilation process. Proton-proton collision is different. But if you are referring to the qq hard scattering I can tell you that the analysis I presented is on low-p_T and not high-p_T physics. In order to study the Drell-Yan process in a clear way you need to look at final states with leptons of opposite charges produced in strong interactions. That kind of physics is completely different from the one we are doing.

DISCUSSION 5   (Scientific Secretaries: G. D'Ali, G. Sartorelli)

- *D'HOKER:*

I would like to make a comment on the problem of the observation of the glueball. The glueball is bound to be very difficult to observe, for two reasons. Firstly, the glueball state mixes with many states of bound quarks. Secondly, at very high energy, a deconfining phase transition occurs and gluons are effectively free and the glueball does not exist. This means that it will always be confused with radiative corrections.

- *KARLINER:*

It is not clear that gluons arrive at thermal equilibrium during the scattering process, so I am not sure how meaningful the concept of temperature is here. However, even if we were to assume that the temperature is so high that it causes a deconfining phase transition, this would probably have some other consequences which would not agree with what one calculates in QCD, assuming zero temperature, and we know that these zero temperature predictions have been experimentally confirmed.

- *ZICHICHI:*

I must remind you of the important message given to us by Professors Dirac, Wigner and Teller: we should be more modest and not pretend to understand everything at once. Now, the reason why people have taken asymptotic freedom very seriously is due to the

fact that, firstly, one knows that the theory is non-Abelian. This
is the starting point. The discovery that in non-Abelian gauge
theories the coupling constant tends to zero when the energy tends
to infinity was very useful in understanding deep-inelastic scattering phenomena. These rather simple considerations form the basis
of hundreds of papers written by theorists. From this moment on,
the confusion increases because none of these papers has really contributed towards a better or deeper understanding of these two very
simple facts. I do believe in asymptotic freedom as an important
property of non-Abelian gauge forces. It is also corroborated by
experiment. We should be very happy that at least we understood
this point. The glueball is something else; it is much more complicated.

- *KARLINER:*

I absolutely agree -- we should believe in asymptotic freedom.
What I said is that if we take the idea seriously that the temperature is not zero, then we might get into trouble and we know that
experimentally we are not in trouble with asymptotic freedom.

- *D'HOKER:*

The final temperature of deconfining phase transition is not in
contradiction with asymptotic freedom. It is just the same thing.

- *SPIEGELGLAS:*

Are there any "next to leading" particles? If so, what good
reasons do you have for neglecting them?

- *ZICHICHI:*

Once you take out the leading proton, you find that the $d\sigma/dx$
distribution (this is the first quantity measured by us) collapses
into the $d\sigma/dx$ distribution of particles produced in e^+e^- interactions. Consequently there is no reason to believe that there is
another leading particle. To be more clear: there is not a second
leading particle.

So, physically speaking, the question is as follows. You have
some energy E which you want to transform into particles. If this
energy E is not associated with any colour singlet state, then the
energy is shared between the particles according to the e^+e^- $d\sigma/dx$
distribution.

If, on the contrary, as is the case in pp collisions, the
energy E is associated with a colour singlet state, then the proton
will not share its energy equally between the produced particles.
This is the "leading particle phenomenon". So, once you have taken
out the leading particle there is no other leading effect left; in
other words, there is no second leading particle. If such a second
leading particle existed it would have been seen in our analysis.

- *PETERSON:*

How do you determine the jet axis?

- *ZICHICHI:*

There are many ways in which this can be done. One can use the thrust or the spericity axis in exactly the same way as for e^+e^- data. In some cases we use the missing momentum axis because we do tet the leading proton, i.e. $\vec{P}_M = \vec{P}_{inc} - \vec{P}_{lead}$, where $\vec{P}_{inc}$ is the momentum of the incoming proton and $\vec{P}_{lead}$ is the momentum of the leading proton. Anyway there are no big differences in the results using different axes; the effects we find do not depend on a particular choice of the jet axis.

- *GABATHULER:*

QCD is a theory of strong interactions. Can we understand the very good agreement between the e^+e^- and pp data within the framework of QCD?

- *ZICHICHI:*

I think that QCD is going to be the common tool with which to explain the data. In fact, when one analyses pp data using the same criteria as in e^+e^- physics, one finds the same results. This is shown by the comparison between the set of quantities, $d\sigma/dx$, $d\sigma/dp_T$, $\langle n_{ch} \rangle$, planarity, etc., measured in (pp) and (e^+e^-) interactions. So, if e^+e^- data are explained in terms of QCD, pp data must also be explained in terms of QCD.

EUROPEAN ORGANIZATION FOR NUCLEAR RESEARCH

EVIDENCE FOR η' LEADING PRODUCTION
IN GLUON–INDUCED JETS

L. Cifarelli
INFN - Salerno, and University of Salerno, Italy

T. Massam
INFN - Bologna, Italy

D. Migani and A. Zichichi
INFN - Bologna, and University of Bologna, Italy

ABSTRACT

In gluon–induced jets, produced in (e^+e^-) annihilation at LEP and studied using the L3 detector, evidence is found for leading production of η', in contrast with η production. This is the first direct proof of a strong gluonic composition of the η'–meson.

Geneva, 18 July 1997

EVIDENCE FOR η' LEADING PRODUCTION IN GLUON–INDUCED JETS

L. Cifarelli, T. Massam, D. Migani and A. Zichichi

Contents

EVIDENCE FOR η' LEADING PRODUCTION IN GLUON–INDUCED JETS

L. Cifarelli, T. Massam, D. Migani and A. Zichichi

1 — Introduction.

One of the most important predictions of QCD is the existence of multigluon states. It has motivated the intensive search for glueballs, where a lot of work has been done; but with still pending results. The existence of a quasi–stable gluon pair is a basic step in this search and a good candidate for such a pair has been (and is) the η'–meson.

We present here the results of a study aimed at establishing the existence of a specific property of the η'–meson, extremely relevant for QCD: its gluonic nature as revealed by the "leading" effect in gluon–induced jets. In fact, if the η' has an important gluonic component, we should expect to see a typical QCD non-perturbative effect: the leading production which has been observed in all hadronic processes [1], where some conserved quantum numbers go from the initial to the final state [2, 3]. This is exactly the effect which is reported in the present paper. In § 2 we briefly discuss the composition of the η'–meson, starting from its 2γ decay up to its present status. In § 3 we report the observed effect and in § 4 the conclusions.

2 — The η'–meson and its composition: status.

For some time, the η' was called X^0, since its pseudoscalar nature was not established and there were mesonic states needed in the tensor multiplet of the Gell-Mann and Ne'eman SU(3)–flavour multiplet structure: $SU(3)_f$. A meson with spin 2 cannot easily decay into 2γ and in fact the 2γ decay mode of the X^0 had not been observed, even when searched for, down to a branching ratio level several times below that of the 2γ decay mode of the η^0, the well-known pseudoscalar neutral meson. This missing 2γ decay mode of the X^0–

meson prevented the X^0–meson from being considered as an $SU(3)_f$ singlet, the 9th member of the pseudoscalar $(q\bar{q})$ system.

The 2γ decay mode of the X^0 was observed in 1968 [4], and its branching ratio was found to be [5, 6] more than twenty times below the 2γ branching ratio of the neutral η–meson (well known to be the 8th member of the $(q\bar{q})$ pseudoscalar $SU(3)_f$ octet).

The discovery of the 2γ decay mode of the X^0–meson gave a strong support to its pseudoscalar nature. However its composition in terms of a quark-antiquark pair remained unclear. In fact, if a meson is made of a $(q\bar{q})$ pair, since quarks carry electric charges, the 2γ decay must be easily allowed. As mentioned above, the branching ratios of the 2γ decay modes of the two mesons were quite different [7]:

$$\frac{\Gamma (\eta^0 \to \gamma\gamma)}{\Gamma (\eta^0 \to total)} = (39.25 \pm 0.31) \%$$

$$\frac{\Gamma (X^0 \to \gamma\gamma)}{\Gamma (X^0 \to total)} = (1.71 \pm 0.33) \%,$$

and the radiative widths of the three pseudoscalar mesons, $\Gamma (\pi^0 \to \gamma\gamma)$, $\Gamma (\eta^0 \to \gamma\gamma)$ and $\Gamma (X^0 \to \gamma\gamma)$ did not follow the theoretical expectations.

Another (not unrelated) difficulty was the X^0 mass. If the X^0–meson had to follow the Gell-Mann-Okubo (quadratic) mass formula, the mixing angle needed for these two pseudoscalar mesons was very small because the X^0 mass is nearly one GeV, compared with the ~ 0.5 GeV mass of the η^0. This mixing, when compared with the $(\omega$-$\phi)$ mixing, also measured by our group [8] to be large (as expected), was the smallest known [7] in all meson physics.

By now, the pseudoscalar nature of the X^0–meson is accepted and this meson is designated with the symbol η'. The notation now used is:

i) η_8, to indicate the 8th component of the $(q\bar{q})$ content of the pseudoscalar meson $SU(3)_f$ multiplet.

ii) η_1, to indicate the $SU(3)_f$ singlet component of the pseudoscalar

$(q\bar{q})$ system.

These two components are not enough to describe the η' composition. In fact, we think we know the reason why the $(\eta-\eta')$ mixing angle is so anomalously small, namely the large gluonic content of the η' originating from the ABJ anomaly (see Ref. [9] for a recent review).

In QCD, the η and η' have played a decisive role. In the early days there was the so-called "η problem" [10]. The theory appeared to demand a pseudoscalar η as an isosinglet made of non-strange quarks, and an η' as an $(s\bar{s})$ state. Consequently the η mass had to be close to the pion mass and the η' mass had to be related to the K mass. The fact that experiments gave a quite different picture was attributed to the ABJ anomaly [11] by Fritzsch, Gell-Mann and Leutwyler [12], by Kogut and Susskind [13], and finally explained as an instanton effect by G. 't Hooft [14]. Instantons induce a strong coupling between the η' and the 2–gluon state, and give this state a high mass, which may both explain why the total width of the η' [6] is so much bigger than that of the η. A quantitative analysis of this phenomenon was done using $\frac{1}{N}$–expansion ideas [15].

Concerning experiments, during the last few years many attempts have been made to find out the gluonic content of the η', for example via a comparative study of the radiative decays of the (J/ψ) into η and η'. However all methods adopted so far are of indirect nature [16].

In the next chapter we present the first direct evidence for a strong gluonic composition of the η'–meson.

3 — The η' leading production in gluon–induced jets.

In the late seventies, the introduction of the effective energy available when two particles interact [17], led to the discovery of Universality Features [3] in the multiparticle hadronic systems produced, no matter whether the interaction be strong, electromagnetic or weak [2]. One of these features is the "leading" production, observed [2] in interactions which can be either purely hadronic (example: pp), or lepto-hadronic (example: νp or ep DIS) or purely leptonic (example: e^+e^- annihilation).

The leading effect takes place when there is a flow [2] of quantum numbers from the initial to the final state. The leading particle produced in a multihadronic process has its quantum numbers, either completely or partly, carried over from the initial state.

The reason why a single particle appears to be "privileged", i.e. leading — since it carries an amount of energy well above the average energy of all the other particles produced in the same multihadronic event — is due to its intrinsic composition. The particle has a strong link with the initial state, in terms of conserved quantum numbers.

For example, in (e^+e^-) annihilation at PETRA, the leading production of D^* has been observed [18] and this is due to the charm composition of the D^*–meson. One of the charmed quarks produced in (e^+e^-) annihilation, via the two-body process $e^+e^- \rightarrow c\bar{c}$, transfers its quantum numbers to the D^* mesonic state, which consequently has the privilege of carrying a large fraction of the total available energy.

Analogously, if in the initial state there is a gluon pair, and if a particle exists which has a large fraction of gluon pairs in its structure, this particle will have a privileged share of the total available energy, i.e. this particle will be leading, because its quantum numbers originate from the initial state. The remaining energy is carried by the multihadronic system associated with the leading particle. The leading particle is the η' and the multihadronic system is the gluon–induced jet.

Using the CERN (e^+e^-) collider (LEP) and the L3 detector, a study was made [19] of the reaction

$$e^+e^- \rightarrow Z^0 \rightarrow (jet)_q + (jet)_{\bar{q}} + (jet)_g$$

where $\qquad\qquad (jet)_q \equiv$ denotes a quark–induced jet

$\qquad\qquad\qquad\qquad (jet)_{\bar{q}} \equiv$ denotes an antiquark–induced jet

$\qquad\qquad\qquad\qquad (jet)_g \equiv$ denotes a gluon–induced jet

at $(\sqrt{s})_{e^+e^-} = Z^0$ peak $\simeq 91$ GeV.

The method to identify the quark– and gluon–induced jets has been described elsewhere [20] and we will limit ourselves to a few essential points. In the sample of (e^+e^-) events, the multihadron systems containing 3 jets are selected. The lowest energy jet is — by definition — assumed to be gluon–induced.

We then proceed to identify the η–meson via its 2γ decay mode:

$$\eta \rightarrow \gamma\gamma . \qquad\qquad (1)$$

Figure 1 shows the invariant mass of the 2γ system in gluon–induced jets. The η–mass peak is clear.

The η'–meson is identified via its decay mode:

$$\eta' \rightarrow \eta\, \pi^+\pi^- .$$
$$ \mathrel{\hookrightarrow} \gamma_M \gamma_M \qquad\qquad (2)$$

Figure 2 shows the $[\eta(\gamma\gamma)\, \pi^+\pi^-]$ invariant mass in gluon–induced jets. The η'–mass peak is also clear.

Notice that the symbol γ_M in (2) indicates the photon needed for the reconstruction of the η which in turn is needed to reconstruct the η' invariant mass. The symbol "M" stands for "Marked".

For the sample of η events in (1) which are not in the η' sample, we use γs which are not γ_M. Thus the two samples of η and η' are totally independent. We call these two samples, sample (1) and sample (2), respectively.

Due to the experimental conditions of observation, the η–mesons of sample (1) must have an energy cut

$$E_\eta \leq 8 \ \text{GeV} .$$

In fact, for η–mesons with higher energy, the 2γ angular separation would not allow, in the symmetric decay case, the two γs to be in two different BGO counters of the L3 detector. For η–mesons from η' decay, in sample (2), the energy cut is

$$E_\eta \leq 5 \ \text{GeV} ,$$

while for η'–mesons, energy cut is

$$E_{\eta'} \leq 8 \ \text{GeV} .$$

Let us define the fractional energy carried either by the η or the η' as:

$$x_{\eta, \eta'} = \frac{E_{\eta, \eta'}}{E(\text{jet})_g}$$

in the Lab–system, where $E(\text{jet})_g$ includes $E_{\eta, \eta'}$.

The maximum energy cuts for η– and η'–mesons are an experimental bias; in order to have available a wide x–range to study the x–distribution of the mesons, we have imposed the condition

$$E(\text{jet})_g \leq 8 \ \text{GeV} .$$

Figure 3 shows the data in terms of the fractional energy distributions $x_{\eta, \eta'}$, for η and η' produced in gluon–induced jets. The x–range extends up to x = 0.7 due to the jet selection conditions and the energy cuts applied. The data in Fig. 3 are not corrected for the acceptance of the L3 apparatus and the overall efficiency of the analysis.

In Fig. 4 we report the same data after corrections.

The data in Figs. 3 and 4 show the η' leading effect which is not present in the η fractional energy distribution.

Let us compare these data with Monte Carlo (MC) simulations. Figure 5 shows the x–distribution of MC events (Jetset 7.5), where the mass of the final state $[\eta(\gamma\gamma)\pi^+\pi^-]$ is in the η'–mass range, with the initial state not being a genuine η'. In other words, the set of particles $[\eta(\gamma\gamma)\pi^+\pi^-]$ is in the same kinematic range as the η', but it is not originated by an η'–meson. Hence Fig. 5 shows the expected behaviour of the MC background.

Analogously, in Fig. 6 the x–distribution of $(\gamma\gamma)$ MC events (Jetset 7.5) in the η–mass range not being originated by η–meson is reported.

By comparing the η' distributions of Figs. 3 and 5, and the η distributions of Figs. 3 and 6, one can conclude that these MC data support the validity of the leading effect for η' production in gluon–induced jets.

There is a further check which is provided by the experimental data, using the fractional energy distribution of the like–charge states

$$\eta\,\pi^+\pi^+ \quad \text{and} \quad \eta\,\pi^-\pi^-$$

in the same mass range as that of the η', selected exactly as in reaction (2). The results are shown in Fig. 7 where no leading effect is present, thus confirming the existence of the η' leading effect in gluon–induced jets.

In Fig. 8 the two ways (Figs. 5 and 7) to estimate the background for the x–distribution in η' production are reported, in order to allow a direct comparison: i) the MC simulation, and ii) the like-charge experimental data. The good agreement between these two methods to study the background rules out the presence of systematic effects that could have affected the experimental data reported in Fig. 4.

Thus the gluon leading effect is indeed present in the η' production process in gluon–induced jets. This could not be the case if the η' composition was dominated by a $q\bar{q}$ content. In fact, in gluon–induced jets the η

production does not show this strong leading effect because the η structure is actually dominated by the $q\bar{q}$ component.

Further studies, along these lines, will allow a more quantitative determination of the gluonic composition of the η', should better QCD theoretical approaches become available.

4 — Conclusion and Discussion.

We have proved that, in gluon–induced jets, the production of the η'–mesons has a remarkable leading effect and that this effect is not present in the production of the η–mesons. Now, let us discuss the consequences.

First consequence: the leading effect is present in gluon–induced jets. Never before has this effect been observed in gluon physics, although it is present in all QCD processes when a set of conserved quantum numbers flows from the initial state to the final state [2]. The observation of the leading effect in gluon jets provides this part of QCD phenomenology (so poorly investigated because of experimental difficulties) with a property which was missing, despite being present in the well investigated part of QCD, i.e. the quark–induced jets.

Concerning the lack of a leading effect in η production, this proves that the two pseudoscalar mesons, η and η', must have drastically different compositions. The η' carries the "signature" of the leading effect in gluon jets. This signature corresponds to a flow [2] of quantum numbers from the initial to the final state. This flow must be associated with the gluons. The straightforward second consequence is that the η' must have an important gluonic component, which the η does not have.

It is interesting to recall two old questions. First, why is the 2γ branching ratio of the η' 20 times smaller than the 2γ branching ratio of the η? Second, why is the mass of the η' so far away from the value expected from the quadratic version of the Gell-Mann-Okubo mass formula, where only

$(q\bar{q})$ mixing in the pseudoscalar nonet of $SU(3)_f$ is taken into account? Hence the <u>third consequence</u> is that the strong gluonic component in the η' structure is an important contribution towards explaining [15, 21] the above two questions: a naive quark model (with colour) happens to answer the question of the radiative widths in an unconvincing way [22], but not the second question.

To sum up: in gluon–induced jets, a strong leading effect has been observed in η' production but not in η. This proves that the η' has an important gluonic composition. Also the result clarifies the long standing discrepancies in the 2γ branching ratios and in the masses of these two pseudoscalar mesons. The present understanding of the masses of the three pseudoscalar mesons (π^0, η, η') appears to be in terms of two competing effects due to the QCD-vacuum and to the (u d s) quark masses. The non-Abelian nature of QCD makes the QCD-vacuum highly non trivial even without quarks. In a "pure" QCD world (without quarks), there would be a lowest energy pseudoscalar state described by the $G_{\mu\nu}\tilde{G}_{\mu\nu}$ operator [14]. Adding quarks to QCD would produce other pseudoscalar states made of $(q\bar{q})$. These states will be affected by the QCD vacuum, which will be strongly effective with the $SU(3)_f$ singlet η_1. The quark-mass terms, on the other hand, produce competing effects. The net result is that the η' is made of $(u\bar{u} + d\bar{d} + \sqrt{2}\ s\bar{s})$ and the η is orthogonal to this state. In terms of $SU(3)_f$ singlet, the weight of η' is something like thirty times stronger than the η. And this is why the η' is so heavy, compared to the other pseudoscalar mesons. Thus the η appears to be made of $(q\bar{q})$ pairs, with a gluonic component much smaller than that of the η'. This is the reason why there is no leading effect for the η in gluon–induced jets. The very heavy pseudoscalar meson, η', is a state with quantum numbers nearest to the QCD gluonic matter and therefore, in addition to the $(q\bar{q})$ pairs, it has an important gluonic component. This is why the η' shows a pronounced leading effect in gluon–induced jets.

330

5 — References.

[1] M. Basile, G. Cara Romeo, L. Cifarelli, A. Contin, G. D'Alì, P. Di Cesare, B. Esposito, P. Giusti, T. Massam, R. Nania, F. Palmonari, V. Rossi, G. Sartorelli, M. Spinetti, G. Susinno, G. Valenti, L. Votano and A. Zichichi
Nuovo Cimento 66A, 129 (1981).

[2] M. Basile, G. Cara Romeo, L. Cifarelli, A. Contin, G. D'Alì, P. Di Cesare, B. Esposito, P. Giusti, T. Massam, R. Nania, F. Palmonari, V. Rossi, G. Sartorelli, M. Spinetti, G. Susinno, G. Valenti, L. Votano and A. Zichichi
Lettere al Nuovo Cimento 32, 321 (1981).

[3] M. Basile, G. Bonvicini, G. Cara Romeo, L. Cifarelli, A. Contin, M. Curatolo, G. D'Alì, C. Del Papa, B. Esposito, P. Giusti, T. Massam, R. Nania, F. Palmonari, G. Sartorelli, G. Susinno, L. Votano and A. Zichichi
Nuovo Cimento 79A, 1 (1984).

[4] D. Bollini, A. Buhler-Broglin, P. Dalpiaz, T. Massam, F. Navach, F.L. Navarria, M.A. Schneegans and A. Zichichi
Nuovo Cimento 58A, 289 (1968).

[5] M. Basile, D. Bollini, P. Dalpiaz, P.L. Frabetti, T. Massam, F. Navach, F.L. Navarria, M.A. Schneegans and A. Zichichi
Nuclear Physics B33, 29 (1971).

[6] P. Dalpiaz, P.L. Frabetti, T. Massam, F.L. Navarria and A. Zichichi
Physics Letters 42B, 377 (1972).

[7] A. Zichichi
in "Evolution of Particle Physics", Academic Press Inc., New York-London (1970), 299.

[8] D. Bollini, A. Buhler-Broglin, P. Dalpiaz, T. Massam, F. Navach, F.L. Navarria, M.A. Schneegans and A Zichichi
Nuovo Cimento 57A, 404 (1968).
see also:
D. Bollini, A. Buhler-Broglin, P. Dalpiaz, T. Massam, F. Navach, F.L. Navarria, M.A. Schneegans and A. Zichichi
Nuovo Cimento 56A, 1173 (1968).

[9] G. Veneziano in "Vacuum and Vacua: The Physics of Nothing", World Scientific, Vol. 33, 16, Erice School 1995, A. Zichichi Ed. and references therein.

[10]   S. Weinberg
       Phys. Rev. D11, 3583 (1975).

[11]   S. Adler
       Phys. Rev. 177, 2426 (1969).
       J. S. Bell and R. Jackiw
       Nuovo Cimento 604A, 47 (1969).

[12]   H. Fritzsch, M. Gell-Mann and H. Leutwyler
       Physics Letters 47B, 365 (1973).

[13]   J. Kogut and L. Susskind
       Phys. Rev. D4, 3501 (1974); D10, 3468 (1974); D11, 3594 (1975).

[14]   G. 't Hooft
       Physical Review Letters 37, 8 (1976); *Physics Report* 142, 357 (1986) and
       references therein.

[15]   E. Witten, *Nuclear Physics* B156, 269 (1979). G. Veneziano, *Nuclear Physics*
       B159, 213 (1979).

[16]   See for example: P. Ball, J.M. Frère and M. Tytgat
       Physics Letters 365B, 367 (1996) and references therein.

[17]   M. Basile, G. Cara Romeo, L. Cifarelli, A. Contin, G. D'Alì, P. Di Cesare,
       B. Esposito, P. Giusti, T. Massam, F. Palmonari, G. Sartorelli, G. Valenti and
       A. Zichichi
       Physics Letters 92B, 367 (1980).

[18]   A. Zichichi, in "Old and New Forces of Nature", Erice School 1985, Plenum
       Press, p. 117.

[19]   D. Migani
       "Prova sperimentale dell'effetto leading in sistemi gluonici", Thesis, Bologna
       University, 18 July 1997.

[20]   M. Acciarri et al., *Physics Letters* B371, 126 (1986).

[21]   G.M. Shore and G. Veneziano, *Nuclear Physics* B381, (1992) 3; 23.

[22]   M.S. Chanowitz
       Physical Review Letters 35, 977 (1975).

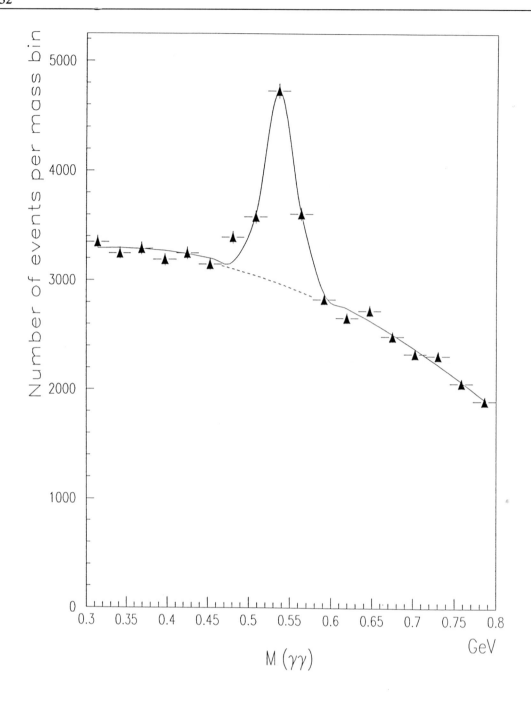

Fig. 1: The (γγ) invariant mass in gluon jets. The η peak is clear.

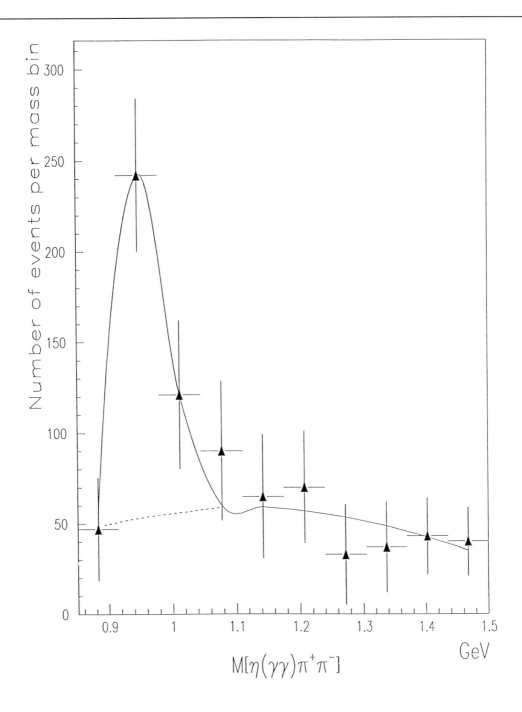

Fig. 2: The $[\eta(\gamma\gamma)\,\pi^+\pi^-]$ invariant mass in gluon jets. The η' peak is clear.

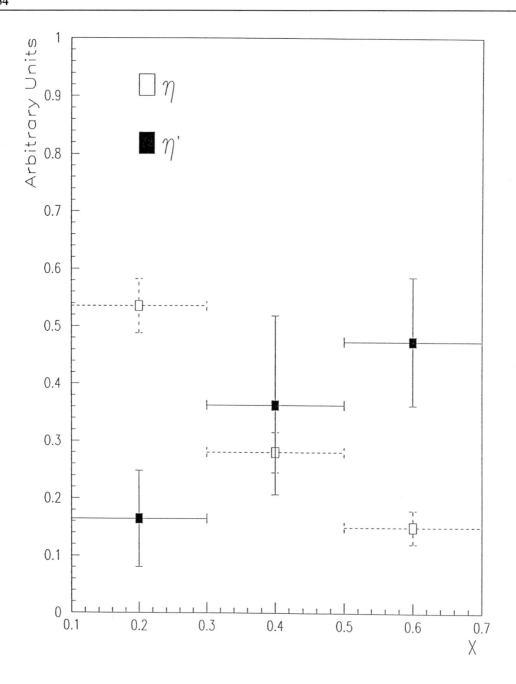

Fig. 3:   x–distributions for η and η' production. Experimental data not
corrected for acceptance and efficiency.

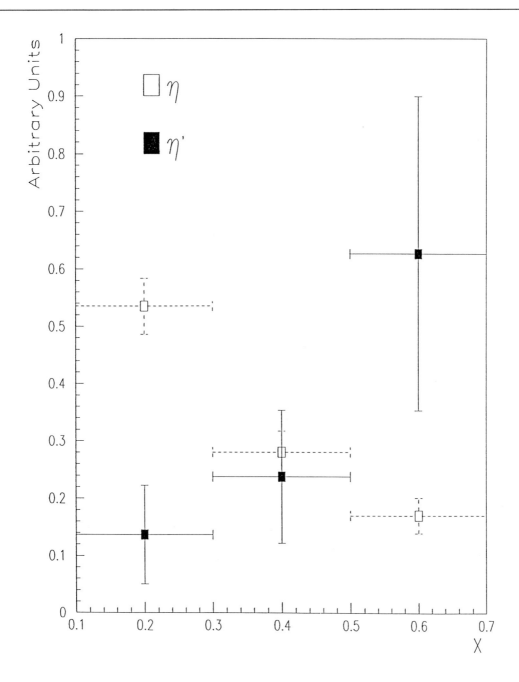

Fig. 4:   x–distributions for η and η ' production. Experimental data
corrected for acceptance and efficiency.

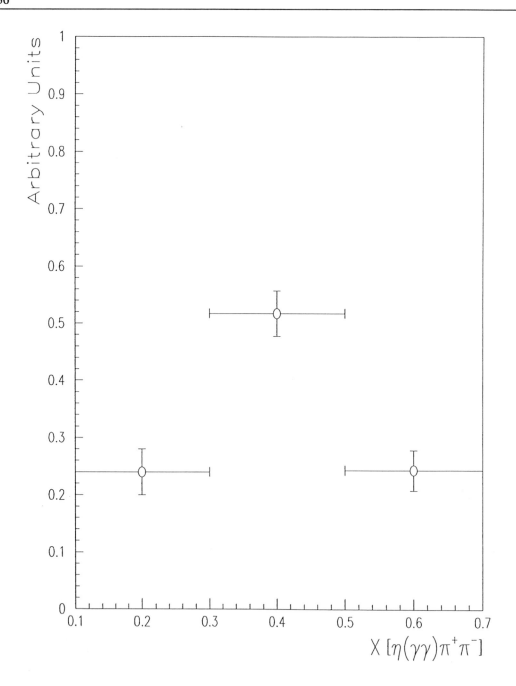

Fig. 5: x–distribution of $[\eta(\gamma\gamma)\,\pi^+\pi^-]$ MC events in the η'–mass range. The final state $[\eta(\gamma\gamma)\,\pi^+\pi^-]$ happens to be in the η'–mass range, while the mother of these events is not an η'–meson.

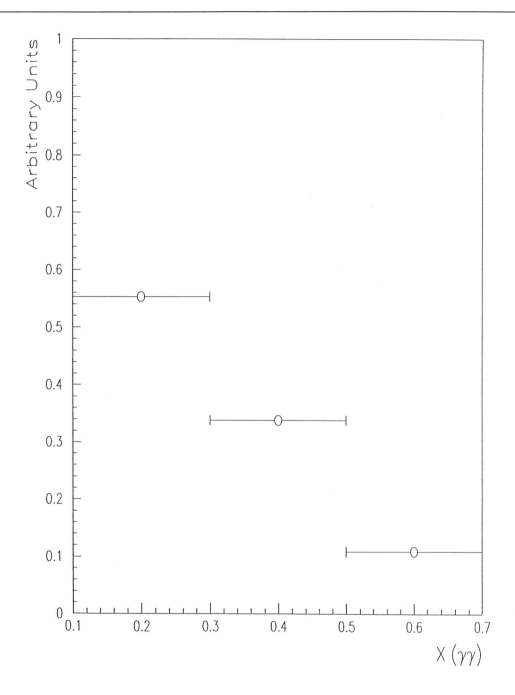

Fig. 6:   x–distribution of (γγ) MC events in the η–mass range. The final state (γγ) happens to be in the η–mass range, while the mother of these events is not an η–meson.

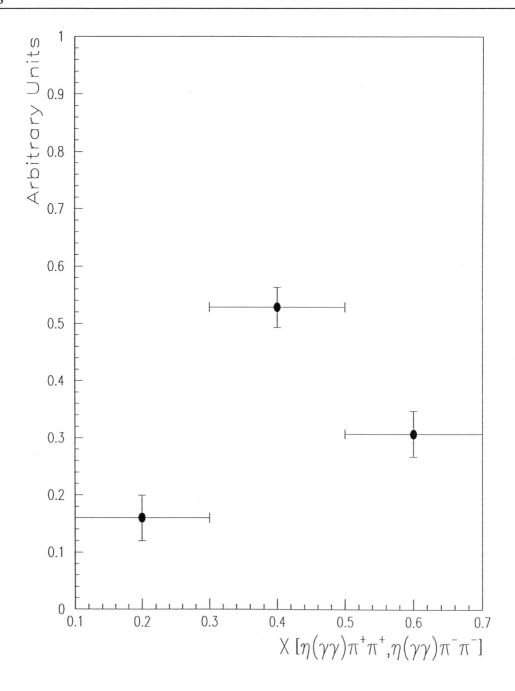

Fig. 7: The experimental x-distribution of like-charge states [η(γγ) π⁺π⁺, η(γγ) π⁻π⁻] whose mass falls in the η'-range.

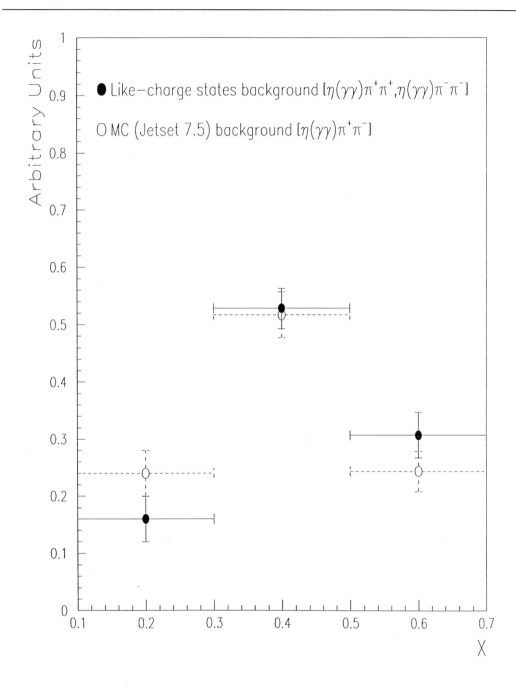

Fig. 8: The background x-distributions of the experimental like–charge states (Fig. 7) and of the MC simulated events (Fig. 5), superimposed.

21

DETAILED REFERENCES OF ALL PAPERS ON THE EFFECTIVE ENERGY AND THE UNIVERSALITY FEATURES

DETAILED REFERENCES OF ALL PAPERS ON
THE EFFECTIVE ENERGY AND THE UNIVERSALITY FEATURES

[1]   EVIDENCE OF THE SAME MULTIPARTICLE PRODUCTION
      MECHANISM IN p-p COLLISIONS AS IN e^+e^- ANNIHILATION
      M. Basile, G. Cara Romeo, L. Cifarelli, A. Contin, G. D'Alì, P. Di Cesare,
      B. Esposito, P. Giusti, T. Massam, F. Palmonari, G. Sartorelli, G. Valenti and
      A. Zichichi
      Physics Letters 92B, 367 (1980).

[2]   THE FRACTIONAL MOMENTUM DISTRIBUTION IN p-p COLLISIONS
      COMPARED WITH e^+e^- ANNIHILATION
      M. Basile, G. Cara Romeo, L. Cifarelli, A. Contin, G. D'Alì, P. Di Cesare,
      B. Esposito, P. Giusti, T. Massam, R. Nania, F. Palmonari, G. Sartorelli,
      G. Valenti and A. Zichichi
      Nuovo Cimento 58A, 193 (1980).

[3]   THE ENERGY DEPENDENCE OF CHARGED PARTICLE
      MULTIPLICITY IN p-p INTERACTIONS
      M. Basile, G. Cara Romeo, L. Cifarelli, A. Contin, G. D'Alì, P. Di Cesare,
      B. Esposito, P. Giusti, T. Massam, R. Nania, F. Palmonari, G. Sartorelli,
      G. Valenti and A. Zichichi
      Physics Letters 95B, 311 (1980).

[4]   STUDY OF THE BASIC PROPERTIES OF LOW-P_T MULTIPARTICLE
      SYSTEMS PRODUCED IN *pp* COLLISIONS REMOVING LEADING
      PROTONS
      M. Basile, G. Cara Romeo, L. Cifarelli, A. Contin, G. D'Alì, P. Di Cesare,
      B. Esposito, P. Giusti, T. Massam, R. Nania, F. Palmonari, G. Sartorelli,
      G. Valenti and A. Zichichi
      Proceedings of the 20th International Conference on "High-Energy Physics",
      Madison, WI, USA, 17-23 July 1980 (AIP, New York, 1981), 86.

[5] MEASUREMENTS OF $\langle p_T^2 \rangle_{in}$ AND $\langle p_T^2 \rangle_{out}$ DISTRIBUTIONS IN HIGH-ENERGY pp INTERACTIONS
M. Basile, G. Cara Romeo, L. Cifarelli, A. Contin, G. D'Alì, P. Di Cesare, B. Esposito, P. Giusti, T. Massam, R. Nania, F. Palmonari, G. Sartorelli, G. Valenti and A. Zichichi
Lettere al Nuovo Cimento <u>29</u>, 491 (1980).

[6] THE RATIO OF CHARGED-TO-TOTAL ENERGY IN HIGH-ENERGY PROTON-PROTON INTERACTIONS
M. Basile, G. Cara Romeo, L. Cifarelli, A. Contin, G. D'Alì, P. Di Cesare, B. Esposito, P. Giusti, T. Massam, R. Nania, F. Palmonari, G. Sartorelli, M. Spinetti, G. Susinno, G. Valenti and A. Zichichi
Physics Letters <u>99B</u>, 247 (1981).

[7] EVIDENCE FOR DOUBLE-JET STRUCTURE IN pp INTERACTIONS AT $\sqrt{s} = 62$ GeV
M. Basile, G. Cara Romeo, L. Cifarelli, A. Contin, G. D'Alì, P. Di Cesare, B. Esposito, P. Giusti, T. Massam, R. Nania, F. Palmonari, G. Sartorelli, M. Spinetti, G. Susinno, G. Valenti, L. Votano and A. Zichichi
Lettere al Nuovo Cimento <u>30</u>, 389 (1981).

[8] THE LEADING EFFECT IN Λ_c^- PRODUCTION AT $\sqrt{s} = 62$ GeV IN PROTON-PROTON COLLISIONS
M. Basile, G. Cara Romeo, L. Cifarelli, A. Contin, G. D'Alì, P. Di Cesare, B. Esposito, P. Giusti, T. Massam, F. Palmonari, G. Sartorelli, G. Valenti and A. Zichichi
Lettere al Nuovo Cimento <u>30</u>, 487 (1981).

[9] THE TRANSVERSE-MOMENTUM DISTRIBUTIONS OF PARTICLES PRODUCED IN pp REACTIONS AND COMPARISON WITH e^+e^-
M. Basile, G. Cara Romeo, L. Cifarelli, A. Contin, G. D'Alì, P. Di Cesare, B. Esposito, P. Giusti, T. Massam, R. Nania, F. Palmonari, G. Sartorelli, M. Spinetti, G. Susinno, G. Valenti, L. Votano and A. Zichichi
Lettere al Nuovo Cimento <u>31</u>, 273 (1981).

[10] CHARGED-PARTICLE MULTIPLICITIES IN *pp* INTERACTIONS AND COMPARISON WITH e^+e^- DATA
M. Basile, G. Cara Romeo, L. Cifarelli, A. Contin, G. D'Alì, P. Di Cesare, B. Esposito, P. Giusti, T. Massam, R. Nania, F. Palmonari, V. Rossi, M. Spinetti, G. Sartorelli, G. Valenti, L. Votano and A. Zichichi
Nuovo Cimento <u>65A</u>, 400 (1981).

[11] THE METHOD OF REMOVING THE LEADING PROTONS IN THE STUDY OF HIGH-ENERGY *pp* REACTIONS, COMPARED WITH THE STANDARD ANALYSIS
M. Basile, G. Cara Romeo, L. Cifarelli, A. Contin, G. D'Alì, P. Di Cesare, B. Esposito, P. Giusti, T. Massam, R. Nania, F. Palmonari, V. Rossi, G. Sartorelli, M. Spinetti, G. Susinno, G. Valenti, L. Votano and A. Zichichi
Nuovo Cimento <u>65A</u>, 414 (1981).

[12] THE INCLUSIVE TRANSVERSE-MOMENTUM DISTRIBUTION IN HADRONIC SYSTEMS PRODUCED IN PROTON-PROTON COLLISIONS
M. Basile, G. Cara Romeo, L. Cifarelli, A. Contin, G. D'Alì, P. Di Cesare, B. Esposito, P. Giusti, T. Massam, R. Nania, F. Palmonari, V. Rossi, G. Sartorelli, M. Spinetti, G. Susinno, G. Valenti, L. Votano and A. Zichichi
Lettere al Nuovo Cimento <u>32</u>, 210 (1981).

[13] THE "LEADING"-PARTICLE EFFECT IN HADRON PHYSICS
M. Basile, G. Cara Romeo, L. Cifarelli, A. Contin, G. D'Alì, P. Di Cesare, B. Esposito, P. Giusti, T. Massam, R. Nania, F. Palmonari, V. Rossi, G. Sartorelli, M. Spinetti, G. Susinno, G. Valenti, L. Votano and A. Zichichi
Nuovo Cimento <u>66A</u>, 129 (1981).

[14] THE "LEADING"-BARYON EFFECT IN STRONG, WEAK, AND ELECTROMAGNETIC INTERACTIONS
M. Basile, G. Cara Romeo, L. Cifarelli, A. Contin, G. D'Alì, P. Di Cesare, B. Esposito, P. Giusti, T. Massam, R. Nania, F. Palmonari, V. Rossi, G. Sartorelli, M. Spinetti, G. Susinno, G. Valenti, L. Votano and A. Zichichi
Lettere al Nuovo Cimento <u>32</u>, 321 (1981).

[15] WHAT WE CAN LEARN FROM HIGH-ENERGY, SOFT (pp) INTERACTIONS
M. Basile, G. Bonvicini, G. Cara Romeo, L. Cifarelli, A. Contin, M. Curatolo, G. D'Alì, B. Esposito, P. Giusti, T. Massam, R. Nania, F. Palmonari, A. Petrosino, V. Rossi, G. Sartorelli, M. Spinetti, G. Susinno, G. Valenti, L. Votano and A. Zichichi
Proceedings of the XIX Course of the "Ettore Majorana" International School of Subnuclear Physics, Erice, Italy, 31 July-11 August 1981: "The Unity of the Fundamental Interactions" (Plenum Press, New York-London, 1983), 695.

[16] THE INCLUSIVE MOMENTUM DISTRIBUTION IN (pp) REACTIONS, COMPARED WITH LOW-ENERGY (e^+e^-) DATA IN THE RANGE $(\sqrt{s})_{e^+e^-} = (3.0 \div 7.8)$ GeV
M. Basile, C. Cara Romeo, L. Cifarelli, A. Contin, G. D'Alì, P. Di Cesare, B. Esposito, P. Giusti, T. Massam, R. Nania, F. Palmonari, V. Rossi, F. Rohrbach, G. Sartorelli, M. Spinetti, G. Susinno, G. Valenti, L. Votano and A. Zichichi
Nuovo Cimento <u>67A</u>, 53 (1982).

[17] A DETAILED STUDY OF $\langle n_{ch} \rangle$ VERSUS E^{had} AND $m_{1,2}$ AT DIFFERENT $(\sqrt{s})_{pp}$ IN (pp) INTERACTIONS
M. Basile, G. Bonvicini, G. Cara Romeo, L. Cifarelli, A. Contin, M. Curatolo, G. D'Alì, P. Di Cesare, B. Esposito, P. Giusti, T. Massam, R. Nania, F. Palmonari, A. Petrosino, V. Rossi, G. Sartorelli, M. Spinetti, G. Susinno, G. Valenti, L. Votano and A. Zichichi
Nuovo Cimento <u>67A</u>, 244 (1982).

[18] HIGH-ENERGY SOFT (pp) INTERACTIONS COMPARED WITH (e^+e^-) AND DEEP-INELASTIC SCATTERING
M. Basile, G. Bonvicini, G. Cara Romeo, L. Cifarelli, A. Contin, M. Curatolo, G. D'Alì, C. Del Papa, B. Esposito, P. Giusti, T. Massam, R. Nania, F. Palmonari, G. Sartorelli, M. Spinetti, G. Susinno, L. Votano and A. Zichichi
Proceedings of the XX Course of the "Ettore Majorana" International School of Subnuclear Physics, Erice, Italy, 3-14 August 1982: "Gauge Interactions: Theory and Experiment" (Plenum Press, New York-London, 1984), 701.

[19]   EXPERIMENTAL PROOF THAT THE LEADING PROTONS ARE NOT CORRELATED
M. Basile, G. Bonvicini, G. Cara Romeo, L. Cifarelli, A. Contin, M. Curatolo, G. D'Alì, C. Del Papa, B. Esposito, P. Giusti, T. Massam, R. Nania, F. Palmonari, G. Sartorelli, G. Susinno, L. Votano and A. Zichichi
Nuovo Cimento <u>73A</u>, 329 (1983).

[20]   DEEP INELASTIC SCATTERING AND SOFT (pp) INTERACTIONS: A COMPARISON
M. Basile, G. Bonvicini, G. Cara Romeo, L. Cifarelli, A. Contin, M. Curatolo, G. D'Alì, C. Del Papa, B. Esposito, P. Giusti, T. Massam, R. Nania, F. Palmonari, G. Sartorelli, G. Susinno, L. Votano and A. Zichichi
Lettere al Nuovo Cimento <u>36</u>, 303 (1983).

[21]   MEASUREMENT OF FORWARD AND BACKWARD MEAN CHARGED-PARTICLE MULTIPLICITIES IN HIGH-ENERGY (pp) SOFT INTERACTIONS AND COMPARISON WITH HIGH-ENERGY NEUTRINO AND ANTINEUTRINO DEEP INELASTIC SCATTERING
G. Bonvicini, G. Cara Romeo, L. Cifarelli, A. Contin, M. Curatolo, G. D'Alì, C. Del Papa, B. Esposito, P. Giusti, T. Massam, R. Nania, G. Sartorelli, G. Susinno, L. Votano and A. Zichichi
Lettere al Nuovo Cimento <u>36</u>, 555 (1983).

[22]   EVIDENCE FOR THE SAME TWO-PARTICLE CORRELATIONS IN RAPIDITY SPACE IN (pp) COLLISIONS AND (e^+e^-) ANNIHILATION
J. Berbiers, G. Bonvicini, G. Cara Romeo, L. Cifarelli, A. Contin, M. Curatolo, G. D'Alì, C. Del Papa, B. Esposito, P. Giusti, T. Massam, R. Nania, G. Natale, F. Palmonari, G. Sartorelli, G. Susinno, L. Votano and A. Zichichi
Lettere al Nuovo Cimento <u>36</u>, 563 (1983).

[23]   STUDIES OF TWO-PARTICLE CORRELATION IN RAPIDITY SPACE IN (pp) COLLISIONS AT $(\sqrt{s})_{pp} = 30, 44$ AND 62 GeV
J. Berbiers, G. Bonvicini, G. Cara Romeo, L. Cifarelli, A. Contin, M. Curatolo, G. D'Alì, C. Del Papa, B. Esposito, P. Giusti, T. Massam, R. Nania, F. Palmonari, G. Sartorelli, G. Susinno, L. Votano and A. Zichichi
Lettere al Nuovo Cimento <u>37</u>, 246 (1983).

[24] EVIDENCE FOR THE SAME INCLUSIVE FRACTIONAL-ENERGY
DISTRIBUTIONS IN SOFT (pp) INTERACTIONS AND IN (μp) DEEP
INELASTIC SCATTERING
G. Bonvicini, G. Cara Romeo, L. Cifarelli, A. Contin, M. Curatolo, G. D'Alì,
C. Del Papa, B. Esposito, P. Giusti, T. Massam, R. Nania, G. Sartorelli,
G. Susinno, L. Votano and A. Zichichi
Lettere al Nuovo Cimento 37, 289 (1983).

[25] HADRON COLLIDERS VERSUS (e^+e^-) COLLIDERS: A
CONTRIBUTION TO THE ROUND TABLE FROM THE BCF GROUP
M. Basile, G. Bonvicini, G. Cara Romeo, L. Cifarelli, A. Contin, M. Curatolo,
G. D'Alì, C. Del Papa, B. Esposito, P. Giusti, T. Massam, R. Nania, G. Natale,
F. Palmonari, G. Sartorelli, M. Spinetti, G. Susinno, L. Votano and A. Zichichi
Proceedings of the 3rd Topical Workshop on "Proton-Antiproton Collider
Physics", Rome, Italy, 12-14 January 1983, CERN 83-04, 409.

[26] STUDY OF SOFT (pp) INTERACTIONS AND COMPARISON WITH
(e^+e^-) AND DIS
M. Basile, G. Bonvicini, G. Cara Romeo, L. Cifarelli, A. Contin, M. Curatolo,
G. D'Alì, C. Del Papa, B. Esposito, P. Giusti, T. Massam, R. Nania,
F. Palmonari, G. Sartorelli, M. Spinetti, G. Susinno, L. Votano and A. Zichichi
Proceedings of the XVIII Rencontre de Moriond on "Gluons and Heavy
Flavours", La Plagne, France, 23-29 January 1983 (Ed. Frontières, Gif-sur-
Yvette, 1983), 175.

[27] SCALE-BREAKING EFFECTS IN (pp) INTERACTIONS AND
COMPARISON WITH (e^+e^-) ANNIHILATIONS
M. Basile, G. Bonvicini, G. Cara Romeo, L. Cifarelli, A. Contin, M. Curatolo,
G. D'Alì, C. Del Papa, B. Esposito, P. Giusti, T. Massam, R. Nania,
F. Palmonari, G. Sartorelli, G. Susinno, L. Votano and A. Zichichi
Lettere al Nuovo Cimento 38, 289 (1983).

[28] THE LEADING EFFECT EXPLAINS THE FORWARD-BACKWARD
MULTIPLICITY CORRELATIONS IN HADRONIC INTERACTIONS
M. Basile, G. Bonvicini, G. Cara Romeo, L. Cifarelli, A. Contin, M. Curatolo,
G. D'Alì, C. Del Papa, B. Esposito, P. Giusti, T. Massam, R. Nania,
F. Palmonari, G. Sartorelli, G. Susinno, L. Votano and A. Zichichi
Lettere al Nuovo Cimento 38, 359 (1983).

[29] TRANSVERSE PROPERTIES OF JETS AT COLLIDER ENERGIES FROM ISR DATA
M. Basile, G. Bonvicini, G. Cara Romeo, L. Cifarelli, A. Contin, M. Curatolo, G. D'Alì C. Del Papa, B. Esposito, P. Giusti, T. Massam, R. Nania, F. Palmonari, G. Sartorelli, G. Susinno, L. Votano and A. Zichichi
Lettere al Nuovo Cimento 38, 367 (1983).

[30] UNIVERSALITY FEATURES IN (pp), (e^+e^-) AND DEEP-INELASTIC-SCATTERING PROCESSES
M. Basile, G. Bonvicini, G. Cara Romeo, L. Cifarelli, A. Contin, M. Curatolo, G. D'Alì, C. Del Papa, B. Esposito, P. Giusti, T. Massam, R. Nania, F. Palmonari, G. Sartorelli, G. Susinno, L. Votano and A. Zichichi
Nuovo Cimento 79A, 1 (1984).

[31] EVIDENCE FOR A WIDER MULTIPLICITY DISTRIBUTION IN GLUON-INDUCED JETS COMPARED WITH QUARK-INDUCED JETS
M. Basile, G. Cara Romeo, L. Cifarelli, A. Contin, G. D'Alì, C. Del Papa, B. Esposito, P. Giusti, T. Massam, R. Nania, F. Palmonari, G. Sartorelli, M. Spinetti, G. Susinno, L. Votano and A. Zichichi
Lettere al Nuovo Cimento 41, 293 (1984).

[32] THE LEADING EFFECT EXPLAINS THE CHARGED-PARTICLE MULTIPLICITY DISTRIBUTIONS OBSERVED AT THE CERN $p\bar{p}$ COLLIDER
M. Basile, G. Cara Romeo, L. Cifarelli, A. Contin, G. D'Alì, C. Del Papa, P. Giusti, T. Massam, R. Nania, F. Palmonari, G. Sartorelli, M. Spinetti, G. Susinno, L. Votano and A. Zichichi
Lettere al Nuovo Cimento 41, 298 (1984).

[33] THE END OF A MYTH: HIGH-P_T PHYSICS
M. Basile, J. Berbiers, G. Cara Romeo, L. Cifarelli, A. Contin, G. D'Alì, C. Del Papa, P. Giusti, T. Massam, R. Nania, F. Palmonari, G. Sartorelli, M. Spinetti, G. Susinno, L. Votano and A. Zichichi
Opening Lecture in Proceedings of the XXII Course of the "Ettore Majorana" International School of Subnuclear Physics, Erice, Italy, 5-15 August 1984: "Quarks, Leptons, and their Constituents" (Plenum Press, New York-London, 1988), 1.

[34] UNIVERSALITY PROPERTIES IN NON-PERTURBATIVE QCD - LEADING D* IN (e^+e^-) AT PETRA
A. Zichichi
Proceedings of the XXIII Course of the "Ettore Majorana" International School of Subnuclear Physics, Erice, Italy, 4-14 August 1985: "Old and New Forces of Nature" (Plenum Press, New York-London, 1988), 117.

[35] (a) LEADING HEAVY FLAVOURED BARYON PRODUCTION AT THE ISR
G. Anzivino, G. Bari, M. Basile, G. Cara Romeo, R. Casaccia, L. Cifarelli, F. Cindolo, A. Contin, G. D'Alì, C. Del Papa, S. De Pasquale, P. Giusti, G. Iacobucci, I. Laakso, G. Maccarrone, T. Massam, R. Nania, F. Palmonari, E. Perotto, G. Prisco, P. Rotelli, G. Sartorelli, G. Susinno, L. Votano, M. Willutzky and A. Zichichi
Proceedings of the 10th Warsaw Symposium on "Elementary Particle Physics", Kazimierzs, Poland, 24-30 May 1987 (Warsaw Univ. and INS, Warsaw, 1987), 237.

(b) NEW RESULTS ON CHARMED AND BEAUTIFUL BARYON PRODUCTION AT THE ISR
G. Anzivino, G. Bari, M. Basile, G. Cara Romeo, R. Casaccia, L. Cifarelli, F. Cindolo, A. Contin, G. D'Alì, C. Del Papa, S. De Pasquale, P. Giusti, G. Iacobucci, I. Laakso, G. Maccarrone, T. Massam, R. Nania, F. Palmonari, E. Perotto, G. Prisco, P. Rotelli, G. Sartorelli, G. Susinno, L. Votano, M. Willutzky and A. Zichichi
Proceedings of the 2nd Topical Seminar on "Heavy Flavours", San Miniato, Italy, 25-29 May 1987, *Nuclear Physics B* (*Proc. Suppl.*) 1B, 55 (1988).

[36] UNIVERSALITY FEATURES IN MULTIHADRONIC FINAL STATES
Old but still interesting results on the properties of multihadronic final states produced in Strong, Electromagnetic and Weak Interactions: a brief recalling
A. Zichichi
Proceedings of the 34th Course of the "Ettore Majorana" International School of Subnuclear Physics, Erice, Italy, 3-12 July 1996: "Effective Theories and Fundamental Interactions" (World Scientific Publishing, 1997), 498.

[37]   EVIDENCE FOR η' LEADING PRODUCTION IN GLUON-INDUCED
        JETS
        Presented by A. Zichichi at the 35th Course of the "Ettore Majorana"
        International School of Subnuclear Physics, Erice, Italy, 26 August-4
        September 1997: "Highlights of Subnuclear Physics: 50 Years Later" (World
        Scientific Publishing, 1999), 474.